&EPA United States
Environmental Protection
Agency

EPA/600/R-13/163 | October 2013
www.epa.gov/ord

Greenspace Restorations

A Framework for Enhancing
Bird Habitat Value of Urban Greenspaces
in the Woonasquatucket Watershed,
Rhode Island, USA

Office of Research and Development
National Health and Environmental Effects Research Laboratory, Atlantic Ecology Division

EPA/600/R-13/163 | October 2013

A framework for enhancing bird habitat value of urban greenspaces in the Woonasquatucket watershed, Rhode Island USA

Richard A. McKinney*
Office of Research and Development
National Health and Environmental Effects Research Laboratory
Atlantic Ecology Division
Narragansett, Rhode Island 02882, USA

Meghan E. Nightingale
Department of Natural Resources Science,
MESM Program
University of Rhode Island
Kingston, RI 02881, USA

*Corresponding author
Phone: (401)782-3133; Fax: (401)782-3030
E-mail address: mckinney.rick@epa.gov

Notice

Abstract

Modifying greenspaces to enhance habitat value has been proposed as a means towards protecting or restoring biodiversity in urban landscapes. In this report, we provide a framework for developing low-cost, low-impact enhancements that can be incorporated during the restoration of greenspaces to enhance their wildlife habitat value. We focus on breeding bird habitat value of urban greenspaces in the Woonasquatucket watershed, a southern New England coastal plain watershed located near Providence, Rhode Island. The report is in two parts: the first is a description of a framework for enhancing bird habitat value of urban greenspaces, and the second describes an empirical study examining bird use of existing greenspaces in the Woonasquatucket watershed. The framework uses existing information on bird-plant associations to provide the elements needed to suggest specific greenspace modifications in terms of plantings that would enhance habitat value for target bird species. Our approach involves i) describing the landscape context of the Woonasquatucket watershed, and, from a bird habitat perspective, identifying advantages and constraints that the surrounding landscape imparts on urban greenspaces in the watershed; ii) identifying a regional bird pool of breeding bird species whose range currently or potentially includes the Woonasquatucket watershed; and iii) identifying a candidate plant list of native woody plant species that support birds in the regional species pool. From these elements a specific target list of bird species can be identified for a restoration of a specific greenspace, which in turn can be used to identify appropriate supporting plants to enhance habitat value. The empirical study investigated bird use of existing greenspace habitats in the Woonasquatucket watershed, and examined links between plant and bird species present at the sites. We surveyed 17 existing greenspaces for breeding birds and woody plant species (trees, shrubs, and vines) during the spring and early summer 2012. Mean bird species richness across all sites was 6.94 ± 0.56 species, and mean abundance was 14.4 ± 8.31 birds. There was a significant positive correlation between bird species richness and the proportion of urban land within 1 km of a site; however, the mean number of human-intolerant species observed was 0.59 ± 0.72 species, suggesting that the increase was a result of an increase in human-tolerant species. Greater than two-thirds of observed bird species had multiple supporting woody plant species present at a site at which they were observed. The mean number of supporting woody plant species per regional bird pool species observed at a site was 3.87 ± 0.26 plants, versus 1.50 ± 0.11 plants for regional bird pool species not observed even though there were supporting plants present for that species. Our results suggest that greenspace restorations that include plantings of multiple supporting plants for a target bird species will have a better chance of attracting the species, and hence increasing bird habitat value. This information may help inform regional resource managers and stakeholders including urban planning departments and local resource conservation organizations involved in planning and carrying out restoration of urban greenspaces.

Key words: greenspace; avian diversity; urban biodiversity; New England; bird-plant associations

Table of Contents

Notice ii

Abstract ii

List of Figures iv

List of Tables iv

I Introduction 1

II The Woonasquatucket Watershed: a bird habitat perspective 4

III Target list of bird species 9

IV Candidate plant species list 14

V Breeding bird use of habitat greenspace 21

Acknowledgements 29

Literature Cited 30

List of Figures

Figure 1-1 Conceptual model for development of suggested plant species....................................3

Figure 2-1 Woonasquatucket River watershed, major rivers and streams...................................5

List of Tables

Table 2-1 Ecological communities known or with the potential to appear in the
Woonasquatucket River watershed 6

Table 3-1 Territory size and Partners in Flight population trend classification 11

Table 4-1 Candidate list of woody plant species proposed to have habitat value 15

Table 5-1 Bird species richness and abundance within 50 m and proportion of urban land 24

Table 5-2 Woody plant species richness and vegetation characteristics measured in 2012.......24

Table 5-3 Comparison of the number of supporting plant species per observed bird species....25

Table 5-4 Mean number of supporting woody plant species per regional bird pool species......26

Table 5-5 Regional bird pool species not observed during 2012 sampling...............................27

I. Introduction

Urban greenspaces include remnant natural lands, areas of ruderal vegetation, parks or nature trails, and vegetated areas created for stormwater management or water quality enhancement. Most cities support the restoration, enhancement, or creation of greenspaces under community development initiatives that promote the integration of built and natural environments. In many areas, urban planners are working to implement stormwater management plans that encourage best management practices such as vegetated buffers, stormwater wetlands, bioretention facilities, and vegetated swales. Urban greenspaces in general are recognized as having many benefits over built environments, but their potential as wildlife habitat is often not realized. This is in part because scientific

Ruderal vegetation in an area that was formerly a city park. Photo: R. McKinney

knowledge about the potential wildlife habitat value of greenspaces is not at the point where it can consistently inform planning and restoration efforts. As a result, management practices that could enhance wildlife habitat of greenspaces are often discounted in the restoration process (Harrison and Davies 2002). With this project we hope to provide information that will facilitate the recognition of the potential habitat value of urban greenspaces, and provide a means by which low-cost, low-impact enhancements can be incorporated during the restoration of greenspaces to enhance their wildlife habitat value. We focus on birds as an indicator species for wildlife habitat value because of their high visibility and positive impacts on the attitudes of urban residents (Bjerke and Ostdahl 2004, Luck et al. 2011), as well as the ready availability of field techniques and modeling approaches to describe their use of urban habitats.

Song Sparrow, *Melospiza melodia*.
Photo: US FWS National Digital Library

The project features a two-phased approach: the first phase focuses on the development of a regional bird pool from which a target list of bird species appropriate to a given restoration project can be identified, as well as a candidate list of native woody plants derived from the habitat requirements of species in the regional bird pool. This phase also includes empirical studies to assess the habitat value of existing greenspaces for breeding birds. Empirical studies include collecting breeding bird abundance and species richness data at urban vegetated patches and greenspaces and using multi-metric modeling to evaluate the effect of habitat and landscape characteristics on bird use of the areas. The second phase of the project used the models developed in phase one to guide the development of a protocol that can be used by planners and restoration managers to optimize an urban greenspace restoration site for bird habitat value (Figure 1-1). The protocol uses habitat requirements of site-specific target bird species to derive a list of appropriate plant species and a landscape plan that, when incorporated into the site design, can help enhance bird habitat value.

This document describes some of the preliminary results of the initial phase of this project, and includes 1) an overview of the Woonasquatucket watershed from a bird's perspective; 2) development of a target list of bird species; 3) identification of a candidate plant species list of native woody plants for use in greenspace restoration; and 4) results of a preliminary study investigating bird use of current greenspace habitats in the watershed. The overall goal of the study is to provide input to support greenspace restoration strategies that include the enhancement of bird habitat value through low-cost, low-impact design practices. We hope this information will be helpful to regional resource managers and stakeholders including urban planning departments, property owners, developers, engineers, consultants, contractors, municipal staff, and local resource conservation organizations involved in planning and carrying out restoration of urban greenspaces. While the target bird species and planting recommendations are specific to the Woonasquatucket watershed, the general principles underlying the development of the various project components will be useful in developing similar recommendations in other urban watersheds.

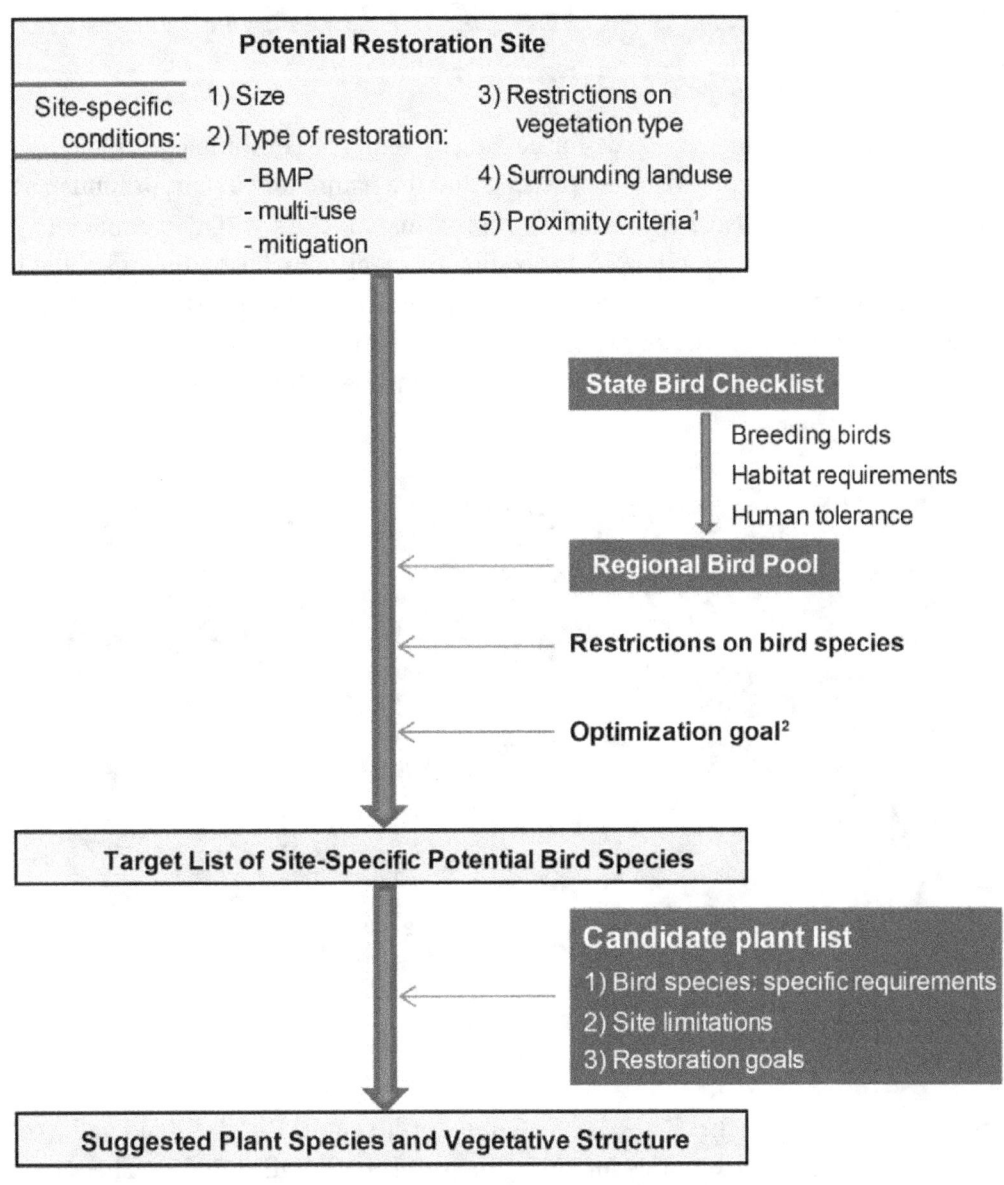

¹Proximity criteria: proximity to existing natural habitats (e.g., wetlands, woodlands), or to known breeding habitat

²Optimize for i) breeding habitat or ii) foraging habitat based on proximity criteria

Figure 1-1. Conceptual model for development of suggested plant species and vegetative structure for greenspace restoration in the Woonasquatucket River watershed, Rhode Island, USA.

II. The Woonasquatucket Watershed: A bird habitat perspective

Use of a specific urban area by birds for nesting or foraging is ultimately driven by the type and arrangement of natural habitats present, and the nature and extent of human activity in the surrounding landscape. A number of proximate factors will also come into play in determining habitat selection; for example, utilization of specific niches within a habitat by a given bird species, and the presence of predators in a given area (Fuller 2012). However, as a starting point it is useful to get a sense of the specific habitat types occurring in the Woonasquatucket watershed and the landscape setting of these habitats with respect to areas of human activity.

Merino Park, a multi-use greenspace adjacent to the Woonasquatucket River in the lower watershed. Photo: R. McKinney

Brief description of the watershed

The 26 km long Woonasquatucket River is located in north-central Rhode Island and a tributary of the Providence River, which empties into Narragansett Bay (RIDEM 2007). The Woonasquatucket flows through six towns of varying urban character, and parts of these towns comprise its 135 km^2 watershed (Figure 2-1). It was formed at the end of the last ice age during the melting of the Laurentide ice sheet that covered much of New England and extended to the south coast of Rhode Island (WRWC 2013). Sand and gravel deposited during melting was scoured by melt water to form the original river channel. In the adjacent floodplain and watershed early-successional wildflowers and fast-growing shrubs helped to establish topsoil, which was subsequently colonized by conifers and ultimately the hardwood forests that dominated prior to human settlement (WRWC 2013). Today the climate associated with the watershed is typical of New England ecoregions: warm, moist summers and cold, snowy winters, with annual average precipitation of about 120 cm yr^{-1} (RIDEM 2007). The watershed is located in the Northeast Coastal Zone ecoregion (Omernik 1987), but perhaps more informative is the ecoregion

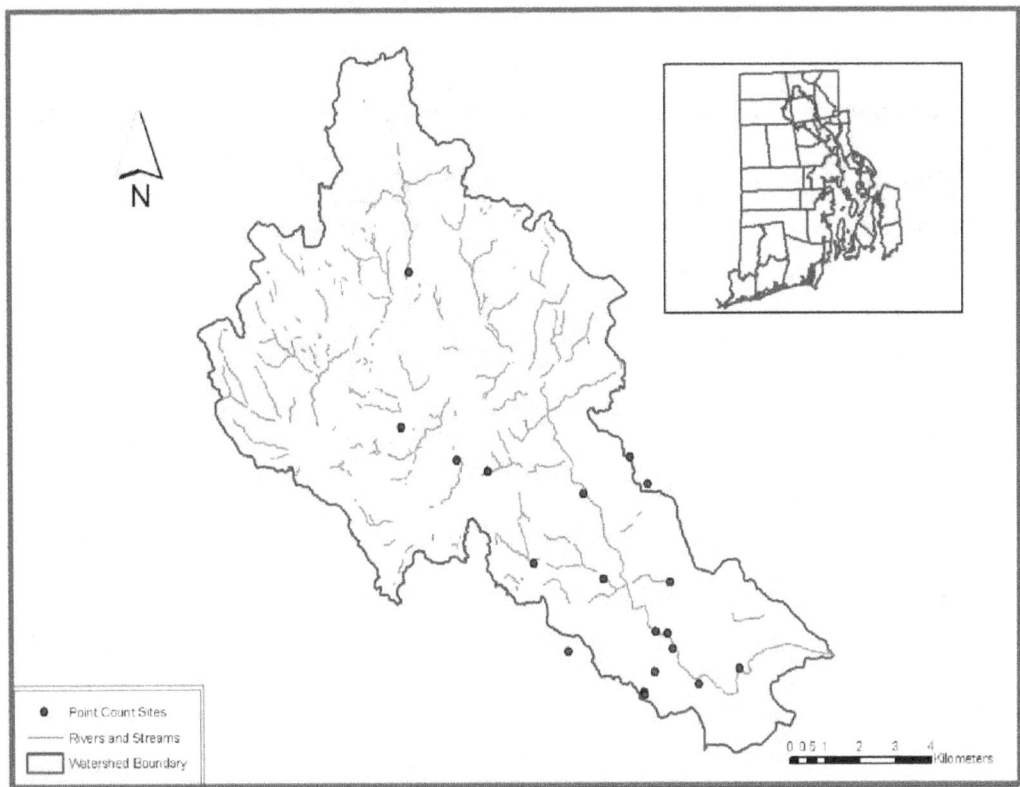

Figure 2-1. Woonasquatucket River watershed, major rivers and streams, and point count sites sampled as part of the empirical study.

classification developed by The Nature Conservancy (Groves et al. 2002), under which the upper watershed is located in the Lower New England Northern Peidmont ecoregion and the lower watershed in the North Atlantic Coast ecoregion. This better describes the somewhat dual nature of the watershed with the upper, more sparsely populated reaches containing a variety of habitat types more characteristic of inland forested habitats and the lower, more densely populated part of the watershed with habitats characteristic of the Atlantic coastal plain.

Ecological communities

A total of 7 upland ecological communities have been identified or potentially appear as small patches in the watershed (Table 2-1). Of these the Mixed Oak / White Pine Forest community is found predominantly in the upper watershed, interspersed with oak and northern hardwood forest communities if present. Sub-dominant tree species in this community will vary by soil type but can include birches (*Betula*), maple (*Acer*), black gum (*Nyssa*), hickory (*Carya*), and American holly (*Ilex*). These communities often have an associated mixed shrub understory, with species again dependent upon soil type and canopy composition. These communities in pristine condition and within an undisturbed landscape setting will provide habitat for forest-dwelling birds including wood warblers, vireos, tanagers, thrushes, woodpeckers, nuthatches, and chickadees. However, upland forest communities in the Woonasquatucket, while often in pristine condition, may be too small to provide significant habitat for many of these species, particularly

Table 2-1. Ecological communities known or with the potential to appear in the Woonasquatucket River watershed, Rhode Island, USA. From Enser et al. (2011)

Upland	Palustrine
Oak forest	Emergent Marsh
Northern Hardwood Forest	Wet Meadow
Hemlock Hardwood Forest	Shrub Swamp
Mixed Oak / White Pine Forest	Northern Peatlands
Ruderal Forest	Forested Swamp
Ruderal Grassland / Shrubland	Seeps, Springs, Bogs
Urban / Recreational Grasses	

those that require large areas of core forest habitat. The relative proximity of these areas to urban land may also limit the number of species that will utilize upland forested habitats (Blair and Johnson 2008). Ruderal communities can be found throughout the watershed including Ruderal Forests and Ruderal Grassland / Shrubland. Ruderal forests formed through succession on areas previously cleared for human activity consist of a variety of early-successional woody plant species including red maple (*Acer*), cherry (*Prunus*), white pine (*Pinus*), red cedar (*Juniperus*), birch (*Betula*), and sassafras (*Sassafras*; Enser et al. 2011). This community can exhibit a diversity of plant species and as a result may provide significant habitat for a variety of urban adapter bird species, given their somewhat less restrictive patch size and landscape setting requirements. Similarly, Ruderal Grassland / Shrubland communities, which include old agricultural fields, clearcuts, hedgerows, and utility rights of way, may harbor a number of urban adapter species. Urban and Recreational Grasses communities, found primarily in the lower watershed, also have the potential to provide habitat for some urban adapters and urban exploiters that can take advantage of the proximity to supplemental food resources (e.g., feeders, seeded lawns, discarded food).

The Woonasquatucket watershed contains about 1400 ha of wetlands, including several ponds, lakes, and reservoirs (RIDEM 2007). Dispersed throughout the watershed are a variety of palustrine wetland

Ruderal grassland/shrubland plant community in a utility right-of-way. Photo: R. McKinney

communities including Emergent Marshes, Wet Meadows, Shrub Swamps, Forested Swamps, Floodplain Forests, and Modified / Managed Marshes (Enser et al. 2011). In addition there may be scattered occurrences of the Seeps, Springs, Vernal Pools community. These wetlands will provide potential habitat for wetland obligate bird species; however, the same caveats apply as in the case of forested habitats: community patch size and landscape setting may be the predominant determinant of which if any of these species will utilize wetland habitats. Riparian habitats in the urban areas of the Woonasquatucket (e.g., Floodplain Forests) may provide enhanced resources for some urban adapter and urban exploiter species, and may be highly utilized despite their size and degree of fragmentation (McKinney et al. 2011). Modified / Managed Marsh communities in the Woonasquatucket are most commonly represented by ruderal marshes, or created wetlands and retention ponds designed to receive stormwater diverted along major roads and industrial areas. They are often dominated by non-native vegetation (e.g., *Phragmites australis, Lythrum salicaria*) and as such are thought to have limited wildlife habitat value; however, in urban settings several urban adapter species, including red-winged blackbirds and song sparrows, may utilize these wetlands as breeding habitat (McKinney and Paton 2009).

Human activity and land use

The Woonasquatucket River's steep descent and narrow width resulted in swift-flowing waters that were harnessed to provide power to industry in the early to mid nineteenth century (Greenwood 2013). A number of factories and mills were built and the river played a pivotal role in the onset of the American Industrial Revolution. Along with industrialization came extensive attempts at water management through the building of dams and reservoirs which changed the landscape in the upper watershed. The Woonasquatucket watershed also became a center for human population growth as the region shifted from an agricultural to industrial based economy (WRWC 2013). This ultimately led to the development of several densely-populated urban areas, particularly in the lower watershed.

Today the watershed remains a mixture of urban and natural land cover. The northwest portion is dominated by wooded hills interspersed with low-density housing (Millar 2004). This part of the watershed contains many natural vegetative communities with the potential to be exploited by urban avoider bird species that are intolerant of human activity. In the upper to mid watershed are a series of narrow valleys that drain into the river; this area is somewhat more heavily populated with areas of medium-density housing in the towns of Johnston and Smithfield. A number of natural areas and small wetlands are found here; in areas removed from human activity they may be used by urban avoider along with urban adapter species. The lower watershed is dominated by dense human settlement and industrial activity in the towns of Providence and North Providence. However, despite the urban character of the lower watershed, there still exist a number of remnant natural areas and urban greenspaces that can be utilized by urban adapter species. Many of these areas are being targeted for acquisition and restoration which could help to preserve and enhance their value as bird habitat (Millar 2004).

Several other considerations are worth noting at this point. First, despite the urban character of much of the mid and lower watershed, the river itself, along with associated tributaries and riparian areas, provides a backbone of natural habitat that can be utilized by birds for breeding and foraging, and can function as movement corridors between patches of fragmented habitat (Millar 2004). Their close proximity to wetlands and the river will presumably allow for continued protection from development through existing state wetlands regulations. Second, while construction of several major highways in the watershed (notably US Route 6 and Interstate Route 95) resulted in fragmentation of traditional neighborhoods and led to negative socio-economic impacts, residual lands from these projects have evolved into significant natural areas in parts of the watershed (WRWC 1998). Several of these areas, particularly those created by the construction of State Route 6, are fairly large, somewhat secluded and inaccessible, and are in relatively close proximity to the river, all of which make them ideal habitat for many urban adapter bird species and possibly some urban avoiders. Many of the areas already have a diversity of plant species that support a variety of wildlife (WRWC 1998). These areas are part of the right-of-way for the roads (i.e., buffer areas around the roads within which development is prohibited), and as a result should escape development pressure. Finally, trends over the past 20 to 30 years have diminished the economic importance of the region leading to plant closures and job dislocation, which in turn contributes to neighborhood destabilization. This has led to vacant factory buildings that can be destroyed by fire and vandalism, and the resulting vacant lands are interspersed

The lower Woonasquatucket watershed has a variety of natural habitats interspersed with built areas: a) Several areas have been restored as multi-use greenspaces.

b) The Woonasquatucket River flows under US Rte 6.

c) The river is surrounded by a vegetated buffer even in commercial and residential areas.

with residential areas (WRWC 1998). Many communities, along with the Woonasquatucket River Watershed Council and a number of other organizations, are working very hard to maintain and restore the integrity of neighborhoods and to incorporate these remnant natural lands into community redevelopment plans (Millar 2004). These areas, along with other urban greenspaces, are currently being utilized by many bird and wildlife species and hold the potential after restoration to contribute to a significant increase in biodiversity in the watershed.

d) High-density residential areas include some natural vegetation and interspersed greenspaces. Photos: R. McKinney

III. Target list of bird species

A key step in developing guidelines for incorporating low-cost, low-impact enhancements to increase bird habitat value of restored greenspaces is to be selective with the species of birds that a restoration will attempt to support. It is well established that many bird species will not utilize habitats in urban landscapes, regardless of condition or in many cases even size of the habitat (Blair 1996, Marzluff et al. 2001, Chace and Walsh 2006). In order to conserve resources during the restoration it's important to exclude these urban avoider species from consideration, and focus on those species that, given the right habitat conditions, will potentially utilize urban greenspaces. Equally important is to exclude species that, although their reported range may encompass the watershed, are not currently found in the region. Some species have undergone historic range contraction or alteration, some with the potential to inhabit the region may not because the required habitat types are not present. The approach taken in this study is to first identify a regional pool of bird species based on these considerations, and from that to develop site-specific target lists of bird species.

A target list of bird species will help identify required habitat characteristics that can be addressed in a landscape plan to enhance bird habitat value as part of a restoration project. In order to minimize the cost of proposed habitat enhancements, the target list should be small (on the order of 10 or fewer species) and specific to a site. Depending on the location of the site in the watershed, the target list of bird species may vary; for example, a target list for

Although fairly common to the region, the yellow-billed cuckoo, *Cuccyzus americanus*, will tend to avoid urban areas. Photo: US FWS National Digital Library

a site near a river or stream will include birds that inhabit riparian areas, whereas an inland restoration site list may consist of birds that primarily utilize upland habitats such as shrubland or urban forest. To identify birds to be considered for inclusion in a site-specific target list, we developed a regional bird pool for the Woonasquatucket watershed. The pool included species that breed in southern New England and, based on life history traits and the ecology of the species, could potentially utilize restored greenspace habitats in the watershed.

Developing a regional bird pool for the Woonasquatucket watershed

Our starting point was the AviBase list of bird species observed in Rhode Island (LePage 2013). AviBase includes a series of species checklists for specific locations throughout the world compiled from a variety of sources including state, regional, and national sightings databases. In North America, as in many other locations, the checklists are regularly updated by local experts in the scientific community and citizen scientists with extensive local experience and expertise gained through bird-watching and other activities. As a first step, the AviBase Rhode Island checklist of 426 species was refined to include only those whose breeding range includes the Woonasquatucket watershed. Using range maps included in avian species accounts in the Birds of North America Online database (Poole 2005), the AviBase list was reduced to 164 breeding species. We then carried out 4 additional refinements to arrive at the target list. First we used data describing the breeding habitats of each species (Ehrlich et al. 1988) to eliminate those species not known to breed in habitats found in the Woonasquatucket: freshwater wetlands, riparian areas, open fields, forests and woodlands, shrublands, and cultivated land.

Gray catbirds, *Dumetella carolinensi,s* are common breeders in the watershed. Globally this species is not considered threatened due to its large range and numbers, however there is some conservation concern in North America because of loss of early successional and grassland habitat. Photo: USFWS National Digital Library

Next we eliminated species classified as urban avoiders according to the urban bird guild classification system proposed by Shwartz et al. (2008). This eliminated species unlikely to inhabit restored urban greenspaces and other fragmented natural areas characteristic of urban landscapes. In the next step we eliminated species that are obligate users of open water habitat, and finally we eliminated rarely-occurring species and those that are considered to be noxious or pest species (e.g., Common Pigeon *Columba livia*). The resulting regional bird pool consisted of 40 species (Table 3-1) from which a site-specific target list of birds can be chosen.

Table 3-1. Territory size and Partners In Flight Population Trend classification of regional bird pool species for the Woonasquatucket River watershed, Rhode Island, USA

Scientific name	Common name	Territory size (ha)[1]	Reference	PIF PT-c[5]
Buteo jamaicensis	red-tailed hawk	425	p	1
Meleagris gallopavo	wild turkey	IN	h	1
Zenaida macroura	mourning dove	IN	l	2
Bubo virginianus	great horned owl	212	p	2
Chaetura pelagica	chimney swift	NT	g	5
Archilochus colubris	ruby-throated hummingbird	IN[2]	o	1
Dryocopus pileatus	pileated woodpecker	< 3.14	e	1
Melanerpes carolinus	red-bellied woodpecker	8.80	s	2
Picoides villosus	hairy woodpecker	1.05	a	1
Picoides pubescens	downy woodpecker	5.10	v	3
Myiarchus crinitus	great crested flycatcher	2.40	t	2
Sayornis phoebe	eastern phoebe	1.77	i	2
Empidonax traillii	willow flycatcher	1.09	r	4
Empidonax minimus	least flycatcher	0.18	p	4
Hirundo rustica	barn swallow	NN	d	4
Petrochelidon pyrrhonota	cliff swallow	NN	c	2
Progne subis	purple martin	NN	u	2
Cyanocitta cristata	blue jay	NT	k	4
Corvus brachyrhynchos	American crow	1.25	f	2
Corvus ossifragus	fish crow	IN[3]	b	2
Poecile atricapillus	black-capped chickadee	3.30	k	1
Baeolophus bicolor	tufted titmouse	4.20	n	2
Sitta carolinensis	white-breasted nuthatch	20.0	k	1
Thryothorus ludovicianus	Carolina wren	0.12	p	1
Troglodytes aedon	house wren	0.40	p	2
Mimus polyglottos	northern mockingbird	0.40	p	4
Dumetella carolinensis	gray catbird	0.11	p	2
Turdus migratorius	American robin	0.12	p	2
Sialia sialis	eastern bluebird	1.01	p	1
Vireo olivaceus	red-eyed vireo	0.73	p	1
Setophaga petechia	yellow warbler	0.04	k	2
Passer domesticus	house sparrow	NT	k	5
Agelaius phoeniceus	red-winged blackbird	0.29	q	4
Quiscalus quiscula	common grackle	NN	m	4
Cardinalis cardinalis	northern cardinal	0.15	p	2
Carpodacus mexicanus	house finch	NT	k	2
Spinus tristis	American goldfinch	IN[4]	j	2
Pipilo erythrophthalmus	eastern towhee	1.90	k	4
Spizella passerina	chipping sparrow	0.60	p	3
Melospiza melodia	song sparrow	0.16	p	4

[1] NT = non-territorial; NN = only territorial in immediate area around the nest; IN = indeterminate

[2] depending on food resources available can range from 0.07 - 3000 ha

[3] nests colonially or semi-colonially

Table 3-1 Continued

[4] varies with type of nesting habitat and nest location
[a] Allison 1947
[b] Bent 1946
[c] Brown & Brown 1995
[d] Brown & Brown 1999
[e] Bull & Jackson 2011
[f] Caffrey 1992
[g] Cink & Collins 2002
[h] Eaton 1992
[i] Hill & Gates 1988
[j] McGraw & Middleton 2009
[k] McKernan & Hartvigsen 2001
[l] Otis et al. 2008
[m] Peer & Bollinger 1997
[n] Pielou 1957
[o] Robinson et al. 1996
[p] Schoener 1968
[q] Searcy & Yakusawa 1995
[r] Sedgewick 2000
[s] Shackleford et al. 2000
[t] Stewart & Robbins 1958
[u] Tarof & Brown 2013
[v] Twomey 1945

[5]Partners in Flight Population Trend descriptions (Panjabi et al. 2012):
1 = Significant large increase (pop'n change > 50%; $P < 0.1$)
2 = Significant small increase or stable (pop'n change 0% to 50%; $P < 0.1$)
3 = Uncertain pop'n change, stable, or possible small decrease ($P > 0.33$; unreliable trend)
4 = Moderate decrease, possible large decrease (pop'n change -15% to -50%; $0.1 < P < 0.33$)
5 = Significant large decrease (pop'n change < -50%; $P < 0.1$)

Factors to consider when developing a target list of bird species for a site

The regional bird pool can be used as a basis for developing a target list of bird species appropriate for a specific restoration site. The target list should be determined in part by any existing conditions at the restoration site (Figure 1-1). Conditions include i) size of the site; ii) the type of restoration planned; iii) any restrictions on vegetation at the site; iv) land use surrounding the site; and v) proximity of the site to existing natural habitats or known breeding habitat.

Size of the site

Site size is important in that many bird species have specific area requirements for breeding territories, or defended areas used for mating, nesting, and from which food is gathered to feed young. Territory size can range from less than a meter for some colonial-nesting species to ten to several hundred hectares for birds of prey (Nice 1941). A recent review of territory size for forest-dwelling passerines listed territory sizes ranging from 0.5 to 6.5 ha (Whitaker and Warkentin

2010). This study also focuses on passerines although not strictly forest-dwelling species but those that utilize urban environments; territory size may be smaller in these birds because of enhanced availability of food resources (Emlen 1974). In spite of this, there may be potential greenspace restoration sites that are too small for certain species. Where available, territory sizes of birds included in the regional bird pool are included in Table 3-1.

Type of restoration and restrictions on vegetation

Type of restoration and restrictions of vegetation types at a site may impact what bird species are feasible to include in the target list. For example, a common goal of greenspace restoration is to enhance stormwater retention in order to meet water quality criteria. Restoring areas as wet vegetated treatment systems, infiltration practices, filtering systems, green roofs, or open channel practices will help meet this goal (RIDEM 2010). Of these, wet vegetated systems (surface wet stormwater basins that provide water quality treatment primarily in a shallow vegetated permanent pool), green roofs, and open channels (vegetated swales) have specific vegetation requirements that may preclude targeting some bird species. Infiltration practices (areas that facilitate retention of surface water into underlying soils), depending on their design, may have more flexibility in the types of vegetation that can be included, or may simply consist of un-vegetated areas. Filtering systems may consist of structural filters with no associated vegetation, but may also include bioretention ponds that may require specific vegetation types. Common among all these is the need to tailor the target list of bird species to the type and characteristics of the greenspace. Many other types of urban greenspaces are not specifically designed for stormwater retention or water quality enhancement, and these may be a target for greenspace restoration as well. Included are formal parks and gardens, remnant natural areas, green corridors, community gardens, and informal recreational areas. While having specific structural requirements (e.g., urban parks often consist primarily of mowed lawns and managed wooded areas), these areas may present options with regard to specific species of plants that may be included.

Surrounding land use and land cover

Landscape setting, or the mix of surrounding land use and land cover, has been shown to play a role in determining use of a site by bird species (Marzluff et al. 2001, Chace and Walsh 2006, Bierwagen 2008). For example, a primary response noted in numerous studies is the absence of human-intolerant species, or 'urban avoiders', at locations in urban areas (Chace and Walsh 2006, Shwartz et al. 2008). We used an urban bird guild classification system proposed by Shwartz et al. (2008) to eliminate those species from the regional bird pool, hence this factor should not have to be explicitly considered when developing a target bird list for a site from our regional bird pool. However, proximity to other natural and semi-natural areas may be worth considering; for example, close proximity of urban wetlands has been shown to influence bird communities in nearby areas (McKinney et al. 2011). Similarly, if a site is near an area known to support breeding birds of a particular species, it may be prudent to consider targeting these species and to include plantings that will provide habitat both for foraging and, if practical, nesting.

Once identified, a site-specific target list of bird species can be used to develop a list of specific plant species whose use could potentially enhance the habitat value of a restoration site for target bird species (Figure 1-1). To simplify this process we developed a candidate plant species list for the Woonasquatucket watershed consisting of woody plants with known habitat value for birds on our regional bird pool.

IV.   Candidate Plant Species List

A candidate list of plants to be considered for planting at a specific restoration site was developed on the basis of published bird-plant associations of woody plant species common to New England (DeGraaf 2002). Because a number of the bird species in the regional bird pool are insectivores or rely on insects for at least part of their food, we also incorporated data on the use of plants by Lepidoptera to identify plants that would provide high habitat value if utilized in a restoration (Tallamy and Shropshire 2009). This approach also added value to our candidate list by including woody plant species that support native insect species, an identified conservation concern for southern New England (Tallamy 2007).

Developing a list of potential woody plant species

As a first step, we considered all woody plant species (trees, shrubs, and vines) described by DeGraaf (2002) as having value as food, nesting habitat, or cover for at least one of the bird species in our regional bird pool. This reference describes most of the commonly-available native woody plants and hence forms a good basis for a list of potential plants to be used in a restoration. We augmented this list with 173 species of woody plants shown to provide habitat for at least one species of Lepidoptera (Tallamy and Shropshire 2009). In order to fine-tune this list we: i) developed a bird-Lepidoptera index to rank the species common between the two lists; ii) added a bird use score to emphasize the importance of the ranked plant species for

Downy serviceberry, *Amelanchier arborea*, grows in a variety of habitats, and its fruits are eaten by many bird and mammal species. Photo: Missouri Botanical Garden

birds; and iii) eliminated a number of species through a series of filters, including growing range and non-native status, in order to arrive at a final list of 36 candidate woody plant species (Table 4-1). Described below are the specific steps taken to arrive at the candidate plant species list.

Table 4-1. Candidate list of woody plant species proposed to have habitat value for birds and Lepidoptera in the Woonasquatucket River watershed, Rhode Island

Family	Genus	Bird - Lepidoptera index	Bird use score	Most utilized (by birds) species	Primary bird use	Growing /planting comments	General comments	Included species
Rosaceae	Rubus	226	259	-	fruit (June–Aug)	fast growing, good in places where human presence is minimal	raspberry, blackberry, dewberry	Rubus allegheneisis, R. flagellaris, R. idaeus, R. occidentalis, R. oderatus
Pinaceae	Pinus	223	253	Eastern white pine, pitch pine	nest, seeds	good in sandy soil	red pine seeds preferred food of PISI; Scots pine seeds RECR	Pinus strobus, P. rigida, P. resinosa, P. sylvestris
Rosaceae	Prunus	236	212	pin cherry	fruit (July–Sept)	good in dry disturbed locations		Prunus pensylvanica, P. serotina, P. virginiana
Cornaceae	Cornus	211	241	alternate-leaf dogwood	fruit (July–Sept)	shady, moist areas	grow slowly; red ossier preferred nest site of AMGO	Cornus alternifolia, C. amomum, C. canadensis, C. racemosa, C. sericea
Rosaceae	Amelanchier	214	214	downy serviceberry	fruit (June–Aug)	fruit production greater when grown in full sun		Amelanchier arborea, A. canadensis, A. laevis
Rosaceae	Malus	227	157	common apple	nest	grow rapidly, but mod life span and may need maintenance	common apple preferred nest site of AMRO, GCFL, REVI, others	Malus "Bob White", Malus "Dorothea"
Juglandaceae	Juglans	201	171	butternut	fruit (Sept–Nov)	fast growing, good in open areas and borders	butternut fruit preferred food of BCCH, WBNU, RBWO, CAWR	Juglans cinerea, J. nigra
Fagaceae	Quercus	227	109	-	nuts (Oct)		scarlet oak occ nest site	Quercus alba, Q. coccinea, Q. palustris, Q. rubra, Q. velutina
Caprifoliaceae	Sambucus	190	237	common elder, scarlet elder	fruit (June–Sept)	hardy, grow rapidly, need full sun, good for thickets		Sambucus canadensis, S. racemosa
Fagaceae	Fagus	197	129	American beech	fruit (Sept–Nov)	shade tolerant, slow growing, long lived, produce fruit after 40 yrs		Fagus grandifolia

Table 4-1 Cont'd

Family	Genus	Bird - Lepidoptera index[1]	Bird use score	Most utilized (by birds) species	Primary bird use	Growing /planting comments	General comments	Included species
Myricaceae	Myrica	201	124	northern bayberry	fruit (June–April)	grows in wide range of soils, good for controlling soil erosion; berries persist through winter	preferred nest site of RWBL	Myrica pensylvanica
Ulmaceae	Ulmus	211	98	American elm	nest, seeds	grows to 100 ft; problem with Dutch elm disease	preferred nest site of BAOR	Ulmus americana
Ericaceae	Vaccinium	199	120	lowbush blueberry	fruit (July–Sept)	hardy, slow-growing, good as ground cover		Vaccinium augustifolium, V. corymbosum
Vitaceae	Parthenocissus	175	169	Virginia creeper	fruit (Aug–Feb)	high-climbing vine, spreads rapidly		Parthenocissus quinquefolia
Anacardiaceae	Rhus	187	124	staghorn sumac	fruit (Aug–Sept)	grows well on steep banks and low nutrient soils; hardy, grow rapidly, little or no care		Rhus hirta
Ulmaceae	Celtis	178	121	common hackberry	fruit (Sept–Nov)	only grows among other trees, in alkaline soil, lower branches occur high on trunk	preferred food of thrushes, fruits persist through winter	Celtis occidentalis
Aquifoliaceae	Ilex	172	146	inkberry, common winterberry	fruit (Mar–June)	hardy, slow-growing, tolerates dry, shady, windy locations	fruits persist into winter	Ilex glabra, I. verticillata, I. laevigata
Nyssaceae	Nyssa	164	164	black tupelo	fruit (Aug–Oct)	good for wet sites, lowlands, moderate growth rate	preferred food of woodpeckers	Nyssa sylvatica
Pinaceae	Picea	202	77	white spruce	nest	large areas only	white and Colorado spruce preferred nest site of AMRO, MODO, NOMO, CHSP	Picea glauca, P. pungens, P. rubens
Grossulariaceae	Ribes	182	108	pasture gooseberry	nest, fruit (July–Sept)	grows in shade and poor soil, good for borders, hedges; good understory shrub	alternate hosts for white pine blister rust	Ribes cynosbati, R. americanum

Table 4-1 Cont'd

Family	Genus	Bird - Lepidoptera index	Bird use score	Most utilized (by birds) species	Primary bird use	Growing /planting comments	General comments	Included species
Moraceae	*Morus*	148	297	red mulberry	fruit (July–Aug)	attractive, fast growing, hardy trees 30–60 ft	white mulberry may be preferred if allowed	*Morus rubra, M. alba*
Betulaceae	*Betula*	208	62	paper birch	seeds	grow rapidly, short (80 yr) life span		*Betula papyrifera, B. lenta, B. alleghaniensis, B. populifolia*
Aceraceae	*Acer*	205	63	red maple	nest, seeds	good roadside trees		*Acer rubrum, A. negundo, A. platanoides, A. saccharinum, A. saccharum*
Rosaceae	*Crataegus*	194	78	Washington hawthorn	fruit, nest	attractive border, plant close together; susceptible to rust, blight: keep away from E. red cedar		*Crataegus phaenopyrum, C. crusgalli*
Rosaceae	*Rosa*	188	81	Virginia rose	fruit, cover	most common rose in New England		*R. virginiana, R. carolina*
Caprifoliaceae	*Viburnum*	193	67	nannyberry, arrowoods	fruit (Aug–Oct)	most are understory shrubs, hardy, moderate growing, good as foundation plantings or background screens (nannyberry)	fruits persist through winter so can be good as an emergency food	*Viburnum dentatum, V. lentago*
Ericaceae	*Gaylussacia*	174	86	highbush huckleberry	fruit (July–Sept)	grows to 3 ft., commonly found among blueberries	preferred food for GRCA	*Gaylussacia baccata, G. brachycera, G. dumosa, G. frondosa*
Juglandaceae	*Carya*	164	101	shagbark hickory	fruit (Sept–Dec)	grows to 80 ft, long-lived	varieties developed for high nut production	*Carya ovata, C. glabra, C. alba*
Rosaceae	*Sorbus*	163	94	American mountain-ash	fruit (Aug–Oct)	grows to 40 ft., cool, moist, sunny locations	good choice for small suburban lawns	*Sorbus americana*
Betulaceae	*Alnus*	189	34	speckled alder, smooth alder	seeds, cover	grows well along watercourses, leaves appear early in spring, form thickets	speckled alder seeds imp food for AMGO	*Alnus incana, A. serrulata*

Table 4-1 Cont'd

Family	Genus	Bird - Lepidoptera index	Bird use score	Most utilized (by birds) species	Primary bird use	Growing /planting comments	General comments	Included species
Oleaceae	*Fraxinus*	169	35	green ash	seeds	fast growing, insect resistant, moderately long-lived	good for streets, parks	*Fraxinus americana, F. nigra, F. pennsylvanica*
Salicaceae	*Populus*	181	27	quaking aspen	nest, buds	rapid growing but short-lived	quaking aspen preferred nest site for BAOR	*Populus tremuloides, P. deltoides, P. grandidentata*
Cupressaceae	*Juniperus*	153	60	common juniper	fruit	hardy, slow-growing, colonizes disturbed areas	good for sandy areas, embankments	*Juniperus communis*
Rosaceae	*Spiraea*	165	31	narrowleaf meadowsweet	fruit	hardy, grow rapidly	good for mass plantings and low borders	*Spiraea alba, S. latifolia, S. tomentosa*
Betulaceae	*Carpinus*	148	32	American hornbeam	cover	hardy, slow-growing, relatively short-lived	begins fruiting 15 years after planting	*Carpinus caroliniana*
Caprifoliaceae	*Symphoricarpos*	132	40	coralberry	fruit, cover	adapt well to rigorous conditions, hardy, tolerate air pollution	nectar attracts RTHU, form thickets good for borders and erosion control along steep embankments	*Symphoricarpos albus, S. orbiculatus*

[1] Higher values of the Bird-Lepidoptera Index and Bird Use Score indicate higher conservation value for regional bird pool species.

Bird-Lepidoptera index

The woody plant species from DeGraaf (2002) were cross-referenced with those described as supporting Lepidoptera and ranked according to a bird-lepidoptera index. The index was developed at the genus level; e.g., all species of the genus Pinus (*Pinus strobus, P. rigida, P. resinosa,* and *P. sylvestris*) found in the study area were grouped together and their utility as bird and lepidoptera habitat reflected in the index. To create the index we first ordered the DeGraaf (2002) species from least (*Arctostaphylos* spp., bearberry: 1 species) to most (*Morus* spp., mulberry: 29 species) bird species supported and assigned them a ranking where a higher number reflects more bird species supported. Support is defined as having demonstrated value as food, a nesting site, or cover for a bird species. For example, black-capped chickadee, *Poecile atricapillus,* have been reported to eat eastern white pine, *Pinus strobus,* seeds; eastern white pine therefore supports black-capped chickadee. We then ordered the 173 species described in Tallamy and Shropshire (2009) from least (23 plant species with 1 species supported) to most (*Quercus* spp., oaks: 532 species) Lepidoptera species supported and assigned a ranking as above. Where species occurred on both lists we added the rankings, and arranged the resulting 65 woody plant species in order from highest combined ranking (reflecting the most combined species of birds and lepidoptera supported) to lowest.

Bird use score

The bird use score was developed from information presented in DeGraaf (2002) and Martin et al. (1951). These publications are summaries of published literature listing birds associated with woody plants, and each publication contains categories of the extent to which bird species rely on a given plant. The intent was to include the extent or degree to which plants provide support to different bird species in the ranking process. We focused on the three resource use categories food, nest, and cover, and summarized use extent categories as normal or preferred. Resource and extent categories were assigned numerical values based on presumed importance to birds (order of importance: nest > food > cover; preferred > normal) resulting in 11 possible combinations of nest, food, and cover along with normal or preferred. These combinations, or use scores, ranged in value from cover only (value 1) through preferred nest, food, and cover (value 19). We then summed the use scores for all bird species using a given woody plant species to arrive at a bird use score for the plant (Table 4-1). A higher score indicates that the plant's resources are utilized to a greater extent by birds. Scores ranged from 27 (*Populus* spp.) to 297 (*Morus* spp.).

Witch hazel, *Hamamelis* spp., is a popular ornamental plant used in landscape design, although it has limited wildlife habitat value. Photo: R. McKinney

Candidate plant list

Woody plant species were ordered in two lists according to bird-Lepidoptera index value (highest to lowest) and bird use score (highest to lowest). The ranks for each plant species were summed, and the plants ordered from lowest to highest according to the sum of ranks. We then eliminated 14 plant species that exhibited little or no habitat value to our target bird species (less than 10% of the highest bird use value), and 6 species that are classified as non-native in southern New England. Several other species were eliminated from the list according to criteria that we felt rendered them unsuitable for use in a restoration setting: 2 species were injurious to humans, 3 species their northern range limit did not include the Woonasquatucket watershed, 2 species exhibited inappropriate growing habits, one species had a propensity to be heavily grazed by deer, and one species was known to harbor injurious plant pest species. The resulting list of 36 woody plants comprises the candidate list of plant species (Table 4-1). The 36 plants on the candidate list comprised 63.9% of the total cumulative bird-Lepidoptera index value for all woody plant species, and 72.9% of the total cumulative bird use score.

Summary

Knowledge of the type and arrangement of natural habitats present, along with their setting the surrounding human-dominated landscape, allowed us to develop a regional bird pool comprised of 40 bird species that could potentially utilize appropriate habitats within the watershed and hence could be included as target species for greenspace restorations. The mean territory size of passerine (songbird) species on the list was 1.95 ± 4.41 ha, and about half of these species had territory sizes under 0.5 ha. It would be reasonable to assume that these species may be able to take advantage of suitable habitats in all but the smallest greenspace restorations for nesting. Thirteen of the 40 species had Partners-in-Flight Population Trend values of 3 or greater, an indication of potential conservation concern for these species (Panjabi et al. 2012). Several of these, including least flycatcher (*Empidonax minimus*), northern mockingbird (*Mimus polyglottos*), red-winged blackbird (*Agelaius phoeniceus*), and song sparrow (*Melospiza melodia*) also have relatively small territory requirements of less than 0.5 ha. Efforts to enhance urban greenspace habitats for these species could be viewed as helping towards their conservation in general, and would add further value to a greenspace restoration.

A candidate list of 36 native woody species was also derived as potential species for inclusion in greenspace restoration efforts in the Woonasquatucket watershed; these plants were ranked in order of the extent to which they support bird species identified as part of the regional pool. As a whole, these candidate plant species represented greater than 50% of the habitat value to birds of all woody plants (native, non-native, ornamental) that were originally considered, and are therefore a good representation of plants with relatively high bird habitat value.

As a first step towards validating the target lists, we undertook an empirical study examining breeding bird use of existing greenspace habitats in the Woonasquatucket watershed. Below are reported results of the study and findings relevant to the potential enhancement of bird habitat value of urban greenspaces during restoration.

V. Breeding bird use of greenspace habitats in an urban southern New England watershed

Introduction

Historical studies of urban birds for the most part focused on describing patterns of community composition and distribution and the effects of urbanization and human disturbance on bird communities (Marzluff et al. 2001, Chace and Walsh 2006). More recently studies have begun to examine the relationship between both the types of vegetation present and vegetative structure and habitat use by birds. For example, a study of bird communities in an urban ecosystem in Brazil showed that species richness had a direct relationship with vegetation complexity (Barbosa de Toledo et al. 2012), and bird-habitat relationships in greenspaces in Delhi, India suggested that high vegetative structural diversity may help maintain bird diversity in these urban habitats (Khera et al. 2009). To date most of these studies have focused on landscape-scale effects of broad vegetation patterns; few studies have looked at finer-scale interactions between birds and specific species of plants present in urban habitats, although some have suggested that these factors may be important (Ortega-Alvarez and MacGregor-Fors 2010, Pennington and Blair 2011).

While few ecological studies have examined bird-plant interactions in urban areas, there have been a number of very effective practical publications geared towards landowners linking the natural history requirements of urban tolerant birds to management actions to enhance their use of urban landscapes (e.g., Roth 1998, DeGraaf 2002, Kress 2006, NWF 2013,and BIB 2013). These guides work from established relationships gleaned from avian literature and past studies linking bird species to specific plant species and the height or structure of vegetation. Analysis of these resources allows the identification of plant species that will support specific bird species by providing food and cover for nests. The result of one such analysis is described in an earlier section of this report, where 36 candidate plant species are ranked in order of providing support for species identified in the regional bird pool for the Woonasquatucket watershed. While this analysis was based ultimately on information from the avian literature, it's not clear whether and to what extent these relationships will be upheld in the Woonasquatucket. The objective of this study was to investigate bird use of existing greenspace habitats in the Woonasquatucket watershed, and to examine links between plant and bird species present at the sites.

In addition to areas targeted for restoration, there are a number of existing urban greenspaces in the Woonasquatucket watershed that can be utilized by birds. Most visible are city parks, which often contain multi-use sports or recreational areas. These greenspaces are the most actively managed, and often are dominated by vegetation not conducive to bird use such as mowed grass, pavement, or gravel surfaces. However these areas may have vegetated buffers consisting of a variety of woody plant species that could be utilized by birds, but this use is often tempered by elevated levels of direct human disturbance in close proximity. Remnant natural areas can include patches of urban forest, shrub, or grassland that were never developed, or developed areas that have reverted back to a semi-natural state. Both of these types of greenspaces can be heavily utilized by birds as they often contain a diversity of plant species and often are free from direct human disturbance. However they can also be dominated by non-native vegetation, particularly in developed areas that have reverted back to a semi-natural state, which may inhibit their use by some bird species. Also present in the watershed are green corridors, including a bike path and several relic transportation rights of way, which can provide habitat for birds. Amenity greenspaces include informal recreation areas associated with housing complexes, apartments, and condominiums, and greenspaces in and around these facilities. Because of their close proximity to humans, these areas generally support a limited number of bird species, although some human-tolerant species can take advantage of these resources. Often discounted as sources of bird habitat are allotments and areas set aside for community gardens. These greenspaces are often not directly used by birds during the growing season, but can be important areas for foraging during the fall and winter months. Considerable amounts of urban land are dedicated to use as cemeteries and churchyards, and because of their semi-natural state and relative lack of human disturbance they can provide bird habitat. As in the case of parks and recreational areas, these greenspaces are often intensely managed, this limits their habitat value to birds.

In this study we examined bird use and vegetation characteristics in 17 urban greenspaces in the Woonasquatucket watershed to determine patterns of association between birds and supporting plant species. The chosen sites represented urban parks, remnant natural areas, green corridors, and amenity greenspaces. By verifying known patterns of bird - plant association, the results will aid in the development of guidelines to enhance the bird habitat value of restored greenspaces. Results will provide information about bird - habitat relationships at the individual patch level, which will compliment broader scale approaches towards enhancing urban bird habitat (McCaffrey and Mannan 2012). Ultimately these guidelines will help inform regional resource managers and stakeholders, including urban planning departments and local resource conservation organizations involved in planning and carrying out restoration of urban greenspaces.

Methods

Sites were randomly selected from a pool of potential greenspace restoration sites identified by the Woonasquatucket River Watershed Council (Providence, RI; www.woonasquatucket.org). Since a majority of the sites were located on private land, we moved the sampling location to the nearest greenspace with public access (Figure 2-1). We established point count locations in an accessible area as close as possible to the center of the greenspace, and conducted 10 min point counts at each station from mid-May to the end of June 2012. All birds seen or heard within a 50 m radius were recorded using a dependent-observer approach. Survey teams consisted of a primary observer who noted bird species and abundance, and a secondary observer who recorded data and noted any individuals missed by the primary observer (Nichols et al. 2000, Forcey et al. 2006). All point counts were conducted between 06:00 and 10:00 hours. As a basis for analysis of plant-bird relationships we used the Woonasquatucket watershed regional bird pool developed previously (Table 3-1 in this document).

We classified vegetation within a 50 m radius around each point count location using vegetation formations described in the National Vegetation Classification Standard (NVCS 2008). Also within a 50 m radius around each point count location we identified all woody plant species present by walking the perimeter of the radius, and a series of transects through each 50 m radius area. Woody plant species were classified as supporting species for birds in the regional bird pool using published information about plant-bird relationships (Martin et al. 1951, DeGraaf 2002).

Additional land-use characteristics were quantified using Geographic Information System (GIS) topographic databases. We obtained GIS data (land use and land cover) from Rhode Island Geographic Information System (RIGIS 2013) and processed data using Environmental Systems Research Institute ARC GIS software (Redlands, CA). Land use and land cover data were summarized from 2004 aerial photography (1:24,000 scale) coded to Anderson modified level 3 (Anderson et al. 1976) to 0.1 ha minimum polygon resolution. We used this information to calculate the percent urban land (residential, commercial, institutional, transportation infrastructure, and industrial land) within a 1 km buffer around the each point count station.

Results

Mean bird species richness measured at the 18 sites was 6.94 ± 0.56 species and mean abundance was 14.4 ± 8.31 birds (Table 5-1). The mean number of human-intolerant species (i.e., urban avoiders; Shwartz et al. 2008) observed was 0.59 ± 0.72 species, therefore, birds detected at the sites were predominantly human-tolerant, mostly urban adapters with a few urban exploiters. There was a significant positive correlation between species richness and the proportion of urban land within 1 km ($r = 0.644$, $F = 10.6$, $p = 0.005$) of a site. Bird abundance also increased significantly with increasing proportion of urban land within 1 km ($r = 0.612$, $F = 8.99$, $p = 0.009$) of a site.

Woody plant species richness ranged from 7 to 14 with a mean of 11.9 ± 0.64 species per site (Table 5-2), and $24.3 \pm 2.76\%$ of species observed were non-native. Trees were the dominant woody plant life form at 10 of the sites, followed by shrubs (4 sites). Other plant life forms dominant at the sites were perennials (2 sites) and mowed lawn (1 site). The mean coverage of dominant life forms across the sites was $62.9 \pm 3.63\%$.

Table 5-1. Bird species richness and abundance within 50 m and proportion of urban land within 1 km of 17 study sites in the Woonasquatucket River watershed, Rhode Island, USA

Site	Species richness 0–50m	Abundance 0–50m	Prop URB 1 km[1]
22	8	15	0.65
41	6	9	0.41
43	2	2	0.31
66	5	11	0.32
188	7	11	0.57
198	7	14	0.88
234	7	11	0.21
248	5	6	0.51
258	11	21	0.85
259	7	21	0.93
349	6	10	0.55
379	4	6	0.57
424	10	16	0.88
425	7	13	0.83
427	9	15	0.81
452	9	34	0.67
454	8	30	0.96

[1] Proportion of urban land in a 1 km buffer around the site

Table 5-2. Woody plant species richness (SR) and vegetation characteristics measured in 2012 at 17 study sites in the Woonasquatucket River watershed, Rhode Island, USA

Site	Woody plant SR	Proportion non-native species	Dominant Life Form (DLF)	Percent cover of DLF (%)
22	10	0.40	tree	60
41	15	0.33	shrub	60
43	10	0.10	tree	100
66	7	0.43	shrub	50
188	13	0.15	tree	65
198	13	0.23	tree	70
234	14	0.29	shrub	50
248	10	0.10	tree	60
258	16	0.19	tree	50
259	16	0.19	tree	50
349	9	0.44	tree	60
379	14	0.21	tree	90
424	12	0.25	tree	80
425	8	0.38	perennial	55
427	11	0.18	shrub	40
452	14	0.07	perennial	60
454	11	0.18	mowed	70

A majority (91.3 ± 2.68%) of the bird species observed at the sites had at least one supporting woody plant species present, and greater than two-thirds (73.2 ± 4.94%) had multiple supporting woody plant species present at a site at which they were observed (Table 5-3; supporting plants are defined as having demonstrated value as food, a nesting site, or cover for a bird species).

Table 5-3. Comparison of the number of supporting plant species per observed bird species in 2012 at 18 study sites in the Woonasquatucket River watershed, Rhode Island, USA

Site	Percent of bird species where 1 or more of the plants at the site were supporting plant species	Percent of bird species where 2 or more of the plants at the site are supporting plant species	Total bird species observed
22	100	88	8
41	89	67	6
43	100	100	2
66	70	50	4
188	88	71	7
198	100	100	6
234	100	43	7
248	100	60	5
258	100	73	11
259	100	71	7
349	100	100	5
379	86	50	4
424	100	40	10
425	67	100	6
427	86	67	9
452	80	78	9
454	86	86	7

Across all sites the mean number of supporting woody plant species per regional bird pool species observed at a site was 3.87 ± 0.26 plants (Table 5-4). For regional bird pool species observed at the sites, supporting plants comprised 33.0 ± 2.11% of all the woody plants reported. At some sites regional bird pool species were not observed even though there were supporting plants present for that species; the mean number of supporting woody plant species per regional bird pool species not observed at a site was 1.50 ± 0.11 plants, and the supporting plants present comprised 12.5 ± 0.67% of the woody plants reported. The five most frequently observed regional bird pool species were American robin, *Turdus migratorius* (present at 16 sites), gray catbird, *Dumetella carolinensis* (13 sites), song sparrow, *Melospiza melodia* (12 sites), common grackle, *Quiscalus quiscula* (7 sites), and house sparrow, *Passer domesticus* (7 sites; Appendix 1). Fifteen regional bird pool species were not observed at any of the sites, and an additional six

Table 5-4. Mean number of supporting woody plant species per regional bird pool species in 2012 at 17 study sites in the Woonasquatucket River watershed, Rhode Island, USA

Site	Total plant species	Supporting plant species per regional bird pool species observed at the site		Supporting plant species per regional bird pool species not observed at the site	
		Mean	Percent of all plants	Mean	Percent of all plants
22	10	4.00	40.0	1.45	14.5
41	15	3.50	23.3	1.92	12.8
43	10	6.00	60.0	1.40	14.0
66	7	2.50	35.7	0.64	9.1
188	13	4.43	34.1	1.58	12.2
198	14	4.00	28.6	1.53	10.9
234	14	4.29	30.6	2.58	18.4
248	10	3.20	32.0	1.63	16.3
258	16	4.82	30.1	1.47	9.2
259	16	6.29	39.3	2.10	13.1
349	9	3.40	37.8	1.06	11.7
379	14	4.00	28.6	1.22	8.7
424	12	2.70	22.5	1.90	15.8
425	8	2.83	35.4	0.81	10.1
427	11	3.33	30.3	1.26	11.5
452	14	3.44	24.6	1.44	10.3
454	11	3.14	28.5	1.48	13.5

were only observed at a single site (Appendix 1). The mean number of regional bird pool species not observed at a site even though greater than 33.0% of the woody plants present were supporting plants was 4.35 ± 0.54 species (Table 5-5). Several regional bird pool species (Eastern bluebird, *Sialia sialis*; Northern mockingbird, *Mimus polyglottos*; Eastern towhee, *Pipilo erythrophthalmus*; Mourning dove, *Zenaida macroura*; American Crow, *Corvus brachyrhynchos*) were not observed or observed at only a single site even though greater than 33.0% of the woody plants present were supporting plants at more than a third of the sites (Table 5-5).

Table 5-5. Regional bird pool species not observed during 2012 sampling at 17 sites in the Woonasquatucket River watershed, Rhode Island, USA even though greater than 33.0% of all plant species at the site were supporting plant species: a) list of all regional bird pool species not observed; b) ranked list of regional bird pool species unobserved at two or more sites with supporting plant species

a)

Site	Regional bird pool species with > 33.0% supporting plants at the site but not observed[1]
22	AMCR, DOWO, EABL, EATO, NOMO, RBWO
41	AMRO, EABL, NOCA, NOMO
43	BLJA, EATO, GRCA, MODO, NOCA, TUTI
66	REVI
188	EABL, EATO, GRCA, MODO
198	NOCA, NOMO
234	AMCR, AMGO, BLJA, COGR, EABL, HAWO, MODO, NOCA, NOMO
248	AMGO, COGR, EATO, HAWO, MODO, NOCA, RBWO, TUTI
258	BCCH, EABL
259	AMCR, DOWO, EABL, NOMO
349	EABL, MODO, NOMO
379	EATO, MODO
424	AMCR, AMGO, EABL, MODO, NOMO
425	EABL, NOCA, NOMO, REVI
427	AMCR, EABL, EATO, NOMO
452	BLJA, EATO, NOCA
454	AMCR, DOWO, GRCA, HAWO, NOCA, NOMO, RBWO

b)

Regional bird pool species	Sites with > 33.0% supporting plants present but not observed at a site	Number observed across all sites
EABL	10	0
NOMO	10	1
NOCA	8	6
EATO	7	1
MODO	7	1
AMCR	6	0
AMGO	3	2
BLJA	3	6
DOWO	3	1
GRCA	3	13
HAWO	3	0
RBWO	3	2
REVI	2	1
TUTI	2	2

[1]AMCR = American Crow, *Corvus brachyrhynchos*
AMGO = American goldfinch, *Spinus tristis*
BCCH = Black-capped chickadee, *Poecile atricapillus*
BLJA = Blue jay, *Cyanocitta cristata*
COGR = Common grackle, *Quiscalus quiscula*
DOWO = Downy woodpecker, *Picoides pubescens*
EABL = Eastern bluebird, *Sialia sialis*
EATO = Eastern towhee, *Pipilo erythrophthalmus*
GRCA = Gray catbird, *Dumetella carolinensis*
HAWO = Hairy woodpecker, *Picoides villosus*
MODO = Mourning dove, *Zenaida macroura*
NOCA = Northern cardinal, *Cardinalis cardinalis*
NOMO = Northern mockingbird, *Mimus polyglottos*
RBWO = Red-bellied woodpecker, *Melanerpes carolinus*
REVI = Red-eyed vireo, *Vireo olivaceus*
TUTI = Tufted titmouse, *Baeolophus bicolor*

Discussion

An increase in bird species richness with increasing urbanization around a site has been observed in several urban habitat types including urban wetlands (McKinney et al. 2011), urban forest remnants (Blair and Johnson 2008), and early-successional habitats (Schlossberg et al. 2011). The increase in bird species richness with increasing urbanization across a variety of greenspace types observed in this study is consistent with these trends, and may be a result of human-tolerant bird species taking advantage of increased resources, or a lack of competition from human-intolerant species at sites with more urban character. For example, some human-tolerant birds may be passing up the less urban sites, even though supporting plants are present, to take advantage of abundant resources at more urban sites. Greenspaces in urban landscapes may also offer some protection from avian or mammalian predators, especially those who will avoid areas of human habitation. Regardless of the cause of this increase in species richness, it is important to note that it may come at an ecological cost as human-tolerant bird species are displacing human-intolerant species once resident at urban sites, resulting in a shift in bird community composition. One consequence of this is the potential for a shift in foraging strategies of resident birds, which in turn may have unforeseen consequences on ecosystem structure and function (Blair and Johnson 2008).

We did not observe any relationship between woody plant species richness and the extent of urbanization at our sites; however, almost a quarter of the species present at a site were categorized as non-native species. It is not clear whether this degree of non-native plant species coverage is higher than that found in natural areas, or just reflects a regional level of non-native species occurrence regardless of landscape setting of a site. Equally important is whether woody plant species community composition changes across a gradient of urbanization, as a result of differing susceptibility to human-generated pollutants. Both of these issues are beyond the scope of the present study, but may have implications for bird use of urban greenspace habitats.

Our results suggest that birds may be focusing on woody plant species that have been identified as providing some habitat value to them, and suggest that the presence of multiple supporting woody plant species increases the probability of finding a bird species at a site. This may be a result of birds looking to utilize the sites for multiple uses, for example for foraging and nesting, which may require the presence of multiple plant species to provide those resources. It may also reflect that different plant species provide their resources at different times, and birds utilizing a site may want to assure that resources are available throughout their period of use. Or it may just be the result of birds' need for a certain level of resource availability that can't be provided by a single plant species. Whatever the driving mechanism, it does appear that from a practical standpoint providing multiple supporting plant species at a site will enhance habitat value.

Our observation that some birds known to inhabit the watershed were not present even though there were multiple supporting plant species present may be the result of human-intolerant birds avoiding the more urban sites in our study. Anecdotally it is well known that a number of bird species are less likely to use areas in the presence of human disturbance; scientific studies have verified this for several species (Blair 1996). What is not as well known is the tolerance threshold, if any, for species in the regional bird pool. The degree to which birds will avoid human-dominated areas may be species specific, and may further preclude some species from being considered as targets for enhancement of greenspace bird habitat value. For example, in our study several species were not observed even though there were abundant supporting plants at more than a third of the sites. This may indicate that these are not possibilities as target species for our study area. However, several of these species have been documented to use urban habitats, including northern mockingbird (Stracey and Robinson 2012) and American crow (Bent 1946), although the relative extent of urbanization at sites where they were observed was not clear.

Greater than half of the species in our regional bird pool for the Woonasquatucket were not observed or observed at only a single site. It may therefore be possible, with adequate habitat enhancement, to increase bird diversity by creating greenspaces with vegetation specifically targeting these species. Some particularly promising candidate species are American goldfinch (*Spinus tristis*), downy woodpecker (*Picoides pubescens*), hairy woodpecker (*Picoides villosus*), red-bellied woodpecker (*Melanerpes carolinus*), and red-eyed vireo (*Vireo olivaceus*). All of these species are somewhat human-tolerant and have been observed in urban habitats (McKinney et al. 2011), so may be attracted to the study area if there are more sites with abundant supporting plants. Other studies have suggested that urban bird diversity can be enhanced by management actions in small greenspaces, particularly where vegetation structure can be modified to better reflect foraging height requirements (Shanahan et al. 2011, McCaffrey and Mannan 2012). These actions would also be consistent with providing additional resources for birds that are scarce in urban areas, which has been suggested to help promote avian biodiversity (Evans et al. 2011). These efforts at the local scale will have to be combined with landscape-scale efforts, such as regional planning initiatives (Pennington and Blair 2011), to move towards the goal of enhancing bird diversity in urban habitats.

Acknowledgements
We would like to thank Marissa Mazzotta, Autumn Oczkowski, and Cathy Wigand for providing comments on the manuscript. Kristen DeMoranville assisted with the preparation of Appendix 3. Mention of trade names or commercial products does not constitute endorsement or recommendation. Although the research described in this article has been funded wholly by the U.S. Environmental Protection Agency, it has not been subjected to Agency-level review. Therefore, it does not necessarily reflect the views of the Agency. This is ORD Tracking Number ORD-005195 of the Atlantic Ecology Division, National Health and Environmental Effects Research Laboratory, Office of Research and Development, U.S. Environmental Protection Agency.

VI. Literature Cited

Allison DG. 1947. Bird populations of forest and forest edge in central Illinois. Master's Thesis. Univ. of Illinois, Urbana, IL.

Anderson JR, Hardy EE, Roach JT, and Whitmer RW. 1976. A land use and land cover classification system for use with remote sensor data. Geological Survey Professional Paper 964. U.S. Geological Survey, Washington, D.C., USA.

Barbosa de Toledo M, Donatelli RJ, Batista GT. 2012. Relation between green spaces and bird community structure in an urban area in Southeast Brazil. Urban Ecosystems 15:111–131.

Bent AC. 1946. Life histories of North American jays, crows and titmice. U.S. National Museum Bulletin 191.

BIB 2013. Birds in Backyards: Guidelines for creating bird habitat. Birdlife Australia, http://www.birdsinbackyards.net/Guidelines-Creating-Bird-Habitats (Accessed June 2013).

Bierwagen BG. 2008. Connectivity in urbanizing landscapes: The importance of habitat configuration, urban area size, and dispersal. Urban Ecosystems 10, 29–42.

Bjerke T, Ostdahl T. 2004. Animal-related attitudes and activities in an urban population. Anthrozoos 17:109–129.

Blair RB. 1996. Land use and avian species diversity along an urban gradient. Ecological Applications 6:506–519.

Blair RB, Johnson EM. 2008. Suburban habitats and their role for birds in the urban-rural habitat network: points of local invasion and extinction? Landscape Ecology 23:1157–1169.

Brown CR, Brown MB. 1995. Cliff Swallow (*Petrochelidon pyrrhonota*), The Birds of North America Online (A. Poole, Ed.). Ithaca: Cornell Lab of Ornithology; Retrieved from the Birds of North America Online: http://bna.birds.cornell.edu/bna/species/149

Brown CR, Brown MB. 1999. Barn Swallow (*Hirundo rustica*), The Birds of North America Online (A. Poole, Ed.). Ithaca: Cornell Lab of Ornithology; Retrieved from the Birds of North America Online: http://bna.birds.cornell.edu/bna/species/452 (Accessed June 2013).

Bull EL, Jackson JA. 2011. Pileated Woodpecker (*Dryocopus pileatus*), The Birds of North America Online (A. Poole, Ed.). Ithaca: Cornell Lab of Ornithology; Retrieved from the Birds of North America Online: http://bna.birds.cornell.edu/bna/species/148 (Accessed June 2013).

Caffrey C. 1992. Female-biased delayed dispersal and helping in American Crows. Auk 109:609–619.

Cink CL, Collins CT. 2002. Chimney Swift (*Chaetura pelagica*), The Birds of North America Online (A. Poole, Ed.). Ithaca: Cornell Lab of Ornithology; Retrieved from the Birds of North

America Online: http://bna.birds.cornell.edu/bna/species/646 (Accessed June 2013).

Chace JF, Walsh JJ. 2006. Urban effects on native avifauna: a review. Landscape and Urban Planning 74:46–69.

DeGraff RM. 2002. Trees, shrubs, and vines for attracting birds. Hanover, N.H: University Press of New England.

Eaton SW. 1992. Wild Turkey (*Meleagris gallopavo*), The Birds of North America Online (A. Poole, Ed.). Ithaca: Cornell Lab of Ornithology; Retrieved from the Birds of North America Online: http://bna.birds.cornell.edu/bna/species/022 (Accessed June 2013).

Ehrlich PR, Dobkin DS, Wheye D. 1988. The birder's handbook: a field guide to the natural history of North American Birds. Simon and Schuster Inc., New York, NY.

Emlen JT. 1974. An urban bird community in Tucson, Arizona: derivation, structure, regulation. Condor 76:184–197.

Enser R, Gregg D, Sparks C, August P, Jordan P, Coit J, Raithel C, Tefft B, Payton B, Brown C, LaBash C, Comings S, Ruddock K. 2011. Rhode Island ecological communities classification. Technical Report. Rhode Island Natural History Survey, Kingston, RI, www.rinhs.org (Accessed May 2013).

Evans KL, Chamberlain DE, Hatchwell BJ, Gregory RD, Gaston KJ. 2011. What makes an urban bird? Global Change Biology 17:32–44.

Forcey GM, Anderson JT, Ammer FK, Whitmore RC. 2006. Comparison of two double-observer point-count approaches for estimating breeding bird abundance. Journal of Wildlife Management 70:1674–1681.

Fuller RJ. 2012. The bird and its habitat: an overview of concepts. Pages 4–36 *In* Fuller, RJ, (Ed.) Birds and habitat: Relationships in changing landscapes, Cambridge University Press.

Greenwood RE. 2013. A brief assessment of the historical significance of the Woonasquatucket River valley. Woonasquatucket River Watershed Council, http://www.woonasquatucket.org/history.php (Accessed June 2013).

Groves CR, Jensen DB, Valutis LL, Redford KH, Shaffer ML, Scott JM, Baumgartner JV, Higgins JV, Beck MW, Anderson MG. 2002. Planning for biodiversity conservation: putting conservation science into practice. Bioscience 52:499–512.

Harrison G, and Davies C. 2002. Conserving biodiversity that matters: Practitioners' perspectives on brownfield development and urban nature conservation in London. Journal of Environmental Management 65:95–108.

Hill SR, Gates JE. 1988. Nesting ecology and microhabitat of the Eastern Phoebe in the central Appalachians. American Midlands Naturalist 120:313–324.

Khera N, Mehta V, Sabata BC. 2009. Interrelationship of birds and habitat features in urban greenspaces in Delhi, India. Urban Forestry & Urban Greening 8:187–196.

Kress SW 2006. The Audubon Society guide to attracting birds, 2nd edition. Cornell University Press, Ithaca, New York.

LePage D 2013. AviBase - the world bird database. Bird Life International, http://avibase.bsc-eoc.org/ (Accessed June 2013).

Luck GW, Davidson P, Boxall D, Smallbone L. 2011. Relations between urban bird and plant communities and human well-being and connection to nature. Conservation Biology 25:816–826.

Martin AC, Zim HS, Nelson AL. 1951. American wildlife and plants. New York: Dover Publications.

Marzluff JM, Bowman R, Donnelly R. 2001. A historical perspective on urban bird research: trends, terms, and approaches. Pgs 1–17 In J. M. Marzluff, R. Bowman, and R. Donnelly [EDS.], Avian ecology and conservation in an urbanizing world. Kluwer Academic, Norwell, MA.

McCaffrey RE, Mannan RW. 2012. How scale influences birds' responses to habitat features in urban residential areas. Landscape and Urban Planning 105:274–280.

McGraw KJ, Middleton AL. 2009. American Goldfinch (*Spinus tristis*), The Birds of North America Online (A. Poole, Ed.). Ithaca: Cornell Lab of Ornithology; Retrieved from the Birds of North America Online: http://bna.birds.cornell.edu/bna/species/080 (Accessed June 2013).

McKernan P, Hartvigsen G. 2001. The territory distribution of breeding songbirds in the Roemer Arboretum, Geneseo, NY. SUNY Geneseo Journal of Science and Mathematics 2:7–15.

McKinney RA, Paton PWC. 2009. Breeding birds associated with seasonal pools in the northeastern United States. Journal of Field Ornithology 80:380–386.

McKinney RA, Raposa KB, Cournoyer RM. 2011. Wetlands as habitat in urbanizing landscapes: patterns of bird abundance and occupancy. Landscape and Urban Planning 100:144–152.

Millar S. 2004. Woonasquatucket greenspace protection strategy. Rhode Island Department of Environmental Management, Office of Sustainable Watersheds, Providence, RI, 121 pgs.

Nice MM. 1941. The role of territory in bird life. American Midland Naturalist 26:441–447.

Nichols JD, Hines JE, Sauer R, Fallon FW, Fallon JE, Heglund PJ. 2000. A double-observer approach for estimating detection probability and abundance from point counts. Auk 117:393–408.

NWF. 2013. Create a bird-friendly habitat. National Wildlife Federation, Reston, VA USA, http://www.nwf.org/how-to-help/garden-for-wildlife/gardening-tips/how-to-attract-birds-to-your-garden.aspx (Accessed June 2013).

NVCS. 2008. National Vegetation Classification Standard, Version 2. Vegetation Subcommittee, Federal Geographic Data Committee, February 2008., http://www.fgdc.gov/standards/projects/FGDC-standards-projects/vegetation/NVCS_V2_FINAL_2008-02.pdf (Accessed June 2013).

Omernik JM. 1987. Ecoregions of the conterminous United States. Map (scale 1:7,500,000). Annals of the Association of American Geographers 77:118–125.

Ortega-Alvarez R, MacGregor-Fors I. 2010. What matters most? Relative effect of urban habitat traits and hazards on urban park birds. Ornitologia Neotropical 21:519–533.

Otis DL, Schulz JH, Miller D, Mirarchi RE, Baskett TS. 2008. Mourning Dove (*Zenaida macroura*), The Birds of North America Online (A. Poole, Ed.). Ithaca: Cornell Lab of Ornithology; Retrieved from the Birds of North America Online: http://bna.birds.cornell.edu/bna/species/117 (Accessed June 2013).

Panjabi AO, Blancher PJ, Dettmers R, Rosenberg KV. 2012. The Partners in Flight handbook on species assessment. Partners in Flight Technical Series No. 3. Rocky Mountain Bird Observatory website: http://www.rmbo.org/pubs/downloads/Handbook2012.pdf (Accessed June 2013).

Peer BD, Bollinger EK. 1997. Common Grackle (*Quiscalus quiscula*), The Birds of North America Online (A. Poole, Ed.). Ithaca: Cornell Lab of Ornithology; Retrieved from the Birds of North America Online: http://bna.birds.cornell.edu/bna/species/271 (Accessed June 2013).

Pennington DN, Blair RB. 2011. Habitat selection of breeding riparian birds in an urban environment: untangling the relative importance of biophysical elements and spatial scale. Diversity and Distributions 17:506–518.

Pielou WP. 1957. A life-history study of the Tufted Titmouse, Parus bicolor Linneaus PhD. Michigan State University, East Lansing.

Poole A. 2005. The Birds of North America Online. Ithaca: Cornell Laboratory of Ornithology; Retrieved from The Birds of North America Online database: http://bna.birds.cornell.edu/BNA/ (Accessed June 2013).

RIDEM. 2007. Woonasquatucket River Fecal Coliform Bacteria and Dissolved Metals Total Maximum Daily Loads. Rhode Island Department of Environmental Management, Office of Water Resources, Providence, RI, 121 pgs.

RIDEM. 2010. Rhode Island Stormwater Design and Installation Standards Manual. Rhode Island Department of Environmental Management and Coastal Resources Management Council, Providence, RI, 487 pgs.

RIGIS. 2013. Rhode Island Geographic Information System Data Repository, http://www.edc.uri.edu/rigis/ (Accessed June 2013).

Robinson TR, Sargent RR, Sargent MB. 1996. Ruby-throated Hummingbird (*Archilochus colubris*), The Birds of North America Online (A. Poole, Ed.). Ithaca: Cornell Lab of Ornithology; Retrieved from the Birds of North America Online: http://bna.birds.cornell.edu/bna/species/204

Roth S. 1998. Attracting birds to your backyard. Rodale Press Emmaus, Pennsylvania.

Schoener TW. 1968. Sizes of feeding territories among birds. Ecology 49:123–141.

Schlossberg S, King DI, Chandler RB. 2011. Effects of low-density housing development on shrubland birds in western Massachusetts. Landscape and Urban Planning 103:64–73.

Searcy WA, Yasukawa K. 1995. Polygyny and sexual selection in Red-winged Blackbirds. Princeton Univ. Press, Princeton, NJ.

Sedgwick JA. 2000. Willow Flycatcher (*Empidonax traillii*), The Birds of North America Online (A. Poole, Ed.). Ithaca: Cornell Lab of Ornithology; Retrieved from the Birds of North America Online: http://bna.birds.cornell.edu/bna/species/533 (Accessed June 2013).

Shackelford CE, Brown RE, Conner RN. 2000. Red-bellied Woodpecker (*Melanerpes carolinus*), The Birds of North America Online (A. Poole, Ed.). Ithaca: Cornell Lab of Ornithology; Retrieved from the Birds of North America Online: http://bna.birds.cornell.edu/bna/species/500 (Accessed June 2013).

Shanahan DF, Possingham HP, Martin TG. 2011. Foraging height and landscape context predict the relative abundance of bird species in urban vegetation patches. Austral Ecology 36:944–953.

Shwartz A, Shirley S, Kark S. 2008. How do habitat variability and management regime shape the spatial heterogeneity of birds within a large Mediterranean urban park? Landscape and Urban Planning 84:219–229.

Stewart RE, Robbins CS. 1958. Birds of Maryland and the District of Columbia. Fish and Wildlife Service, North American Fauna No. 62.

Stracey CM, Robinson SK. 2012. Are urban habitats ecological traps for a native songbird? Season-long productivity, apparent survival, and site fidelity in urban and rural habitats. Journal of Avian Biology 43:50–60.

Tallamy DW. 2007. Bringing nature home. Timberland Press, London.

Tallamy DW, Shropshire KJ. 2009. Ranking Lepidopteran use of native versus introduced plants. Conservation Biology 23: 941–947.

Tarof S, Brown CR. 2013. Purple Martin (*Progne subis*), The Birds of North America Online

(A. Poole, Ed.). Ithaca: Cornell Lab of Ornithology; Retrieved from the Birds of North America Online: http://bna.birds.cornell.edu/bna/species/287 (Accessed June 2013).

Twomey AC. 1945. The bird population of an elm-maple forest with special reference to aspection, territorialism and coactions. Ecological Monographs 15:173–205.

Whitaker DM, Warkentin IC. 2010. Spatial ecology of migratory passerines on temperate and boreal forest breeding grounds. Auk, 127:471–484.

WRWC. 1998. Woonasquatucket river greenway plan. Woonasquatucket River Watershed Council, http://www.wwrc.org/documents/Greenway_Master_Plan.pdf (Accessed June 2013).

WRWC. 2013. Woonasquatucket watershed overview. Woonasquatucket River Watershed Council, http://www.woonasquatucket.org/overview.php (Accessed June 2013).

Appendix 1. Presence (1) / absence (0) of bird species* within 0–50m of point count sites during 2012 at 17 study sites in the Woonasquatucket River watershed, Rhode Island, USA

Common name	Scientific name	Site number																
		22	41	43	66	188	198	234	248	258	259	349	379	424	425	427	452	454
Red-tailed Hawk	*Buteo jamaicensis*	0	0	0	0	0	0	0	0	0	0	0	0	0	0	0	0	0
Wild Turkey	*Meleagris gallopavo*	0	0	0	0	0	0	0	0	0	0	0	0	0	0	0	1	0
Mourning Dove	*Zenaida macroura*	0	0	0	0	0	0	0	0	1	0	0	0	0	0	0	0	0
Great Horned Owl	*Bubo virginianus*	0	0	0	0	0	0	0	0	0	0	0	0	0	0	0	0	0
Chimney Swift	*Chaetura pelagica*	0	0	0	0	0	0	0	0	0	0	0	0	0	0	0	0	0
Ruby-throated Hummingbird	*Archilochus colubris*	0	0	0	0	0	0	0	0	0	0	0	0	0	0	0	0	0
Northern Flicker	***Colaptes auratus***	0	0	0	0	0	0	0	0	0	0	0	0	0	0	0	1	1
Pileated Woodpecker	*Dryocopus pileatus*	0	0	0	0	0	0	0	0	0	0	0	0	0	0	0	0	0
Red-bellied Woodpecker	*Melanerpes carolinus*	0	0	0	0	0	0	0	0	1	0	0	1	0	0	0	0	0
Hairy Woodpecker	*Picoides villosus*	0	0	0	0	0	0	0	0	0	0	0	0	0	0	0	0	0
Downy Woodpecker	*Picoides pubescens*	0	0	0	0	0	0	0	0	0	0	0	0	0	0	1	0	0
Eastern Kingbird	***Tyrannus tyrannus***	0	0	0	0	0	0	0	0	0	0	0	0	0	0	0	1	0
Great Crested Flycatcher	*Myiarchus crinitus*	0	0	0	0	0	0	0	0	0	0	0	0	0	0	0	0	0
Eastern Phoebe	*Sayornis phoebe*	0	0	0	0	1	0	0	0	0	0	0	0	0	0	0	0	0
Willow Flycatcher	*Empidonax traillii*	0	0	0	0	0	0	0	0	0	0	0	0	0	0	0	0	0
Least Flycatcher	*Empidonax minimus*	0	0	0	0	0	0	0	1	0	0	0	0	0	0	0	0	0
Barn Swallow	*Hirundo rustica*	1	0	0	0	0	0	0	0	0	0	0	0	0	0	0	0	0
Cliff Swallow	*Petrochelidon pyrrhonota*	0	0	0	0	0	0	0	0	0	0	0	0	0	0	0	0	0
Purple Martin	*Progne subis*	0	0	0	0	0	0	0	0	0	0	0	0	0	0	0	0	0
Blue Jay	*Cyanocitta cristata*	1	0	0	0	0	0	0	1	1	0	0	1	0	1	1	0	0
American Crow	*Corvus brachyrhynchos*	0	0	0	0	0	0	0	0	0	0	0	0	0	0	0	0	0
Fish Crow	*Corvus ossifragus*	0	0	0	0	0	0	0	0	0	0	0	0	0	0	0	0	0
Black-capped Chickadee	*Poecile atricapillus*	0	0	0	0	1	0	0	1	0	0	0	0	1	0	0	0	0
Tufted Titmouse	*Baeolophus bicolor*	0	0	0	0	1	0	0	0	0	0	0	0	0	0	0	1	0
White-breasted Nuthatch	*Sitta carolinensis*	0	0	0	0	0	0	1	0	0	0	0	0	0	0	0	0	0
Carolina Wren	*Thryothorus ludovicianus*	0	0	0	0	0	0	0	0	0	0	0	0	1	0	1	0	0
House Wren	*Troglodytes aedon*	0	0	0	0	0	0	0	0	0	0	0	0	0	0	0	0	0
Northern Mockingbird	*Mimus polyglottos*	0	0	0	0	0	0	0	0	1	0	0	0	0	0	0	0	0

Appendix 1 Cont'd

Common name	Scientific name	22	41	43	66	188	198	234	248	258	259	349	379	424	425	427	452	454
										Site number								
Brown Thrasher	*Toxostoma rufum*	0	0	0	0	0	0	0	0	1	0	0	0	0	0	0	0	0
American Robin	*Turdus migratorius*	1	0	1	1	1	1	1	1	1	1	1	1	1	1	1	1	1
Eastern Bluebird	*Sialia sialis*	0	0	0	0	0	0	0	0	0	0	0	0	0	0	0	0	0
European Starling	*Sturnus vulgaris*	0	0	0	0	0	0	0	0	1	1	0	0	0	0	1	1	1
Red-eyed Vireo	*Vireo olivaceus*	0	0	0	0	0	0	0	0	0	0	0	0	0	0	1	0	0
Black-and-white Warbler	*Mniotilta varia*	0	0	0	0	0	0	0	0	0	0	0	0	1	0	0	0	0
Yellow Warbler	*Setophaga petechia*	0	0	0	0	0	0	1	0	0	0	0	0	1	0	0	0	1
Yellow-rumped Warbler	*Setophaga coronata*	0	0	0	0	0	0	1	0	0	0	0	0	0	0	0	0	0
Prairie Warbler	*Setophaga discolor*	0	1	0	0	0	0	0	0	0	0	0	0	0	0	0	0	0
House Sparrow	*Passer domesticus*	1	0	0	0	1	1	0	0	0	1	1	0	1	1	0	0	1
Red-winged Blackbird	*Agelaius phoeniceus*	0	0	0	0	0	1	1	0	1	0	0	0	0	0	1	1	0
Common Grackle	*Quiscalus quiscula*	1	1	1	0	1	1	0	0	0	0	0	0	1	0	0	1	0
Baltimore Oriole	*Icterus galbula*	1	0	0	0	0	0	0	0	0	1	0	0	0	0	0	0	0
Northern Cardinal	*Cardinalis cardinalis*	1	0	0	0	1	0	0	0	1	1	1	1	0	0	1	0	0
Indigo Bunting	*Passerina cyanea*	0	0	0	0	1	0	0	1	0	0	0	0	0	0	0	0	0
House Finch	*Carpodacus mexicanus*	0	0	0	0	0	0	0	0	0	0	0	0	0	0	0	0	1
American Goldfinch	*Spinus tristis*	0	0	1	0	0	0	0	0	0	0	0	0	0	1	0	0	0
Eastern Towhee	*Pipilo erythrophthalmus*	0	1	0	0	0	0	0	0	0	0	0	0	0	0	0	0	0
Chipping Sparrow	*Spizella passerina*	0	0	0	0	0	0	0	0	0	0	0	0	1	0	0	0	0
Field Sparrow	*Spizella pusilla*	0	1	0	0	0	0	0	0	0	0	0	0	0	0	0	0	0
Song Sparrow	*Melospiza melodia*	0	1	0	1	0	1	1	1	1	0	1	0	1	1	0	1	1

*Species in bold are not included in the regional bird pool.

Appendix 2. Presence (1) / absence (0) of woody plant species within 0–50m of point count sites during 2012 at 17 study sites in the Woonasquatucket River watershed, Rhode Island, USA

Common name	Scientific name	22	41	43	66	188	198	234	248	258	259	349	379	424	425	427	452	454
Norway Spruce	*Picea abies*	0	1	0	0	0	0	0	0	0	0	0	0	0	0	0	0	0
Blue Spruce	*Picea pungens*	1	1	0	0	0	1	0	0	1	0	0	0	0	1	0	0	0
White Pine	*Pinus strobus*	0	1	0	0	0	1	0	0	1	0	0	0	0	0	0	1	0
Pitch Pine	*Pinus rigida*	0	0	0	0	1	1	0	0	1	0	0	0	0	0	1	0	0
Atlantic White Cedar	*Chamaecyparis thyroides*	0	0	0	0	0	0	1	0	0	1	0	1	0	0	0	0	0
Eastern Red Cedar	*Juniperus virginiana*	1	0	0	0	1	0	0	0	0	0	0	0	0	0	0	0	1
Tulip tree	*Liriodendron tulipifera*	0	0	1	0	1	0	1	1	0	0	0	0	0	0	0	0	0
Sassafras	*Sassafras albidium*	0	0	0	0	0	0	0	1	0	0	0	1	1	0	0	1	0
Witch Hazel	*Hamamelsis virginiana*	0	0	0	0	0	0	1	0	0	0	0	0	0	0	0	0	0
American Elm	*Ulmus americana*	0	0	0	0	0	0	0	0	0	0	0	1	0	0	0	0	0
Chinese Elm	*Ulmus parvifolia*	0	1	1	1	1	0	0	0	0	1	0	0	0	0	0	0	0
Mulberry	*Morus alba*	0	1	0	0	0	0	0	0	0	0	0	0	0	0	0	0	0
Black Walnut	*Juglans nigra*	1	0	0	0	0	0	0	0	0	1	0	0	0	1	1	0	0
Sweet Fern	*Comptonia peregrina*	0	0	0	0	0	0	0	0	1	1	0	0	0	0	0	0	0
Sweetgale	*Myrica gale*	0	0	0	0	0	0	1	1	0	1	0	0	0	0	0	1	1
Chestnut	*Castanea dentata*	0	0	0	0	0	1	0	0	0	0	0	0	0	1	0	0	1
White Oak	*Quercus alba*	0	0	0	0	0	0	0	0	1	1	0	0	0	0	1	0	0
Red Oak	*Quercus borealis*	0	0	0	0	0	0	0	0	0	0	0	0	1	0	0	0	0
Black Birch	*Betula lenta*	0	0	0	0	0	0	0	0	0	1	0	0	0	0	0	1	1
River Birch	*Betula nigra*	1	1	1	1	1	1	1	0	0	1	1	0	0	1	0	0	1
Grey Birch	*Betula populifolia*	0	0	0	0	0	0	0	0	0	0	1	0	0	0	0	0	0
Bigtooth Aspen	*Populus grandidentata*	0	0	0	0	1	0	0	0	0	0	0	0	0	0	0	0	0
Quaking Aspen	*Populus tremuloides*	0	0	0	0	0	0	1	0	0	1	1	1	0	0	0	0	1
Black Willow	*Salix nigra*	0	0	0	0	0	1	0	1	1	0	0	0	0	0	0	0	1
Pussy Willow	*Salix discolor*	0	0	0	0	0	0	0	0	0	1	1	0	0	0	0	1	1
Clethra	*Clethra alnifolia*	0	0	0	0	0	1	1	0	0	1	1	1	1	1	1	0	0
Low-bush Blueberry	*Vaccinium angustifolium*	0	0	0	0	0	0	0	0	0	0	1	1	1	0	0	0	0
High-bush Blueberry	*Vaccinium corymbosum*	0	0	0	0	0	0	0	1	1	0	0	0	0	0	0	0	0
Climbing Hydrangea	*Hydrangea anomala*	0	0	0	0	0	1	0	0	1	1	0	0	0	0	1	0	1
Meadowsweet	*Spiraea alba*	0	0	0	0	0	0	1	0	0	0	0	1	1	0	0	0	0

Appendix 2 Cont'd

										Site number								
Common name	Scientific name	22	41	43	66	188	198	234	248	258	259	349	379	424	425	427	452	454
American Red Raspberry	Rubus idaeus	1	0	1	0	0	0	0	0	0	0	0	0	1	0	0	1	0
Blackberry	Rubus allegheniensis	1	0	0	0	0	1	0	0	0	0	0	0	0	0	0	0	0
Multiflora Rose	Rosa multiflora	0	1	0	0	0	0	0	0	0	0	0	0	0	0	0	0	0
Black Cherry	Prunus serotina	0	0	1	0	0	1	1	1	0	0	0	0	0	0	1	1	0
Pear	Pyrus spp.	0	1	0	0	0	0	0	0	0	0	0	0	0	0	0	0	0
Crabapple	Malus spp.	0	1	0	0	0	0	1	0	0	1	1	1	1	1	1	0	0
Wisteria	Wisteria sinensis	1	1	0	1	0	0	0	0	0	1	0	0	0	0	0	0	0
Black Locust	Robinia pseudoacacia	0	1	0	0	0	0	1	0	0	0	0	0	1	0	1	0	0
Autumn Olive	Eleagnus umbellata	0	1	0	0	1	1	1	0	1	0	0	0	1	0	1	0	0
Redtwig Dogwood	Cornus sericea	1	0	0	0	0	0	0	0	0	0	0	0	0	0	0	0	0
Kousa Dogwood	Cornus kousa	0	0	0	0	0	0	0	0	1	0	0	0	0	0	1	1	0
Black Gum	Nyssa sylvatica	0	0	0	0	0	1	0	0	0	0	0	0	0	1	0	1	0
Oriental Bittersweet	Celastrus orbiculatus	1	1	0	0	1	0	0	1	1	0	0	1	0	0	0	0	0
Winterberry	Ilex verticullata	0	0	1	0	0	0	1	0	0	0	0	0	0	0	0	0	0
Virginia Creeper	Parthenocissus quinquefolia	0	0	0	0	0	0	0	0	0	0	0	1	0	0	0	0	0
Grapevine	Vitis spp.	0	0	1	1	0	1	0	0	0	0	0	0	0	0	0	0	0
Norway Maple	Acer platenoides	0	0	0	0	0	0	0	0	0	1	0	1	0	0	0	0	0
Silver Maple	Acer sachharinum	0	0	0	0	1	0	0	0	0	1	0	0	1	1	1	0	0
Red Maple	Acer rubrum	0	0	1	1	1	1	1	1	0	1	1	0	1	0	1	1	1
Boxelder	Acer negundo	0	0	0	0	0	0	0	0	0	0	0	0	0	1	0	0	1
Smooth Sumac	Rhus glabra	0	0	1	1	1	0	1	1	1	1	1	0	0	0	1	0	1
Tree of Heaven	Ailanthus altissima	0	1	1	0	0	0	0	0	1	0	0	1	0	0	0	0	0
Staghorn Sumac	Rhus typhina	0	0	0	0	0	1	0	1	0	0	0	0	0	0	0	1	0
Winged Sumac	Rhus copallinum	0	0	0	0	0	1	0	0	1	0	0	0	0	0	0	0	0
Poison Ivy	Toxicodendron radicans	0	0	0	0	1	0	0	1	0	1	1	0	0	0	0	0	0
Privet	Ligustrum spp.	1	0	1	0	0	0	0	1	1	0	1	0	1	0	0	1	0
Green Ash	Fraxinus pennsylvanica	0	0	1	1	1	0	1	0	1	0	0	1	0	0	0	0	1
Catalpa	Catalpa bignonioides	0	0	0	0	0	0	0	1	1	0	0	0	0	0	0	1	0
Japanese Honeysuckle	Lonicera japonica	0	1	0	0	0	0	0	0	0	0	0	0	0	0	0	1	0
Southern Arrowwood	Viburnum dentatum	0	0	0	1	0	0	0	0	1	1	0	0	1	0	0	0	0
Mapleleaf V burnum	Viburnum acerifolia	0	0	0	0	0	0	0	0	1	1	0	0	0	0	0	0	0
Greenbriar	Smilax rotundifolia	0	0	1	0	1	0	0	0	1	0	0	1	1	0	0	1	0

Appendix 3. Growth requirements and life history characteristics of woody plants either observed during 2012 at 17 study sites in the Woonasquatucket River watershed, Rhode Island, USA, or identified in the candidate plant species list

Species	Observed?	Common name	Invasive/ native	Sun amount	Soil texture	Growth rate	Growth habit	Size class (ft)	Hardiness (RI 5–7)	Lifespan	Commercial	Additional
Acer negundo	Y	boxelder	native	full sun, part shade, full shade	fine, medium, coarse	rapid	tree	35–60	3–8	short	available	does best in riparian zones
Acer rubrum	Y	red maple	native	full sun, part shade	fine, medium, coarse	rapid	tree	35–68	3–9	short	available	does best in wet environments
Acer saccharinum	Y	silver maple	native	full sun, part shade	fine, medium, coarse	rapid	tree	45–95	3–9	moderate	available	looks un-kept if un-pruned; lifts sidewalks; good tree for away from homes
Acer saccharum	Y	sugar maple	native	full sun, part shade, full shade	medium, coarse	slow	tree, shrub	60–80	3–8	long	available	
Alnus incana	Y	gray alder	native	full sun, part shade, full shade	fine, medium, coarse	rapid	tree, shrub, thicket	15–25	2–6	short	available	nitrogen fixing
Alnus serrulata	Y	hazel alder	native	full sun, part shade	fine, medium, coarse	rapid	tree, shrub	12–30	3–8	moderate	available	nitrogen fixing
Amelanchier arborea	Y	common serviceberry	native	full sun, part shade	medium, coarse	slow	tree, shrub	25–36	5–8	moderate	available	used as a street plant-attractive
Amelanchier canadensis	Y	Canadian serviceberry, shadbush, juneberry		full sun, part shade, full shade	fine, medium, coarse	moderate	tree, shrub	20–23	4–10	long	available	found naturally in bogs
Amelanchier laevis	Y	allegheny serviceberry	native	full sun, part shade, full shade	medium, coarse	moderate	tree, shrub	30–35	4–8	short	available	sensitive to drought
Betula alleghaniensis	Y	yellow birch	native	full sun, part shade	fine, medium, coarse	slow	tree	25–75	3–7	moderate	field collections only	usually found in moist soils
Betula lenta	Y	cherry birch, sweet birch	native	full shade, part shade	medium, coarse	moderate	tree	15–60	4–9	moderate	field collections only	

Appendix 3 Cont'd

Species	Observed?	Common name	Invasive/native	Sun amount	Soil texture	Growth rate	Growth habit	Size class (ft)	Hardiness (RI 5-7)	Lifespan	Commercial	Additional
Betula papyrifera	Y	paper birch	native	full sun, part shade, full shade	fine, medium, coarse	rapid	tree	40–70	2–7	moderate	available	
Betula populifolia	Y	gray birch	native	full sun, part shade, full shade	fine, medium, coarse	rapid	tree, thicket	25	3–6	short	available	
Carpinus caroliniana	N	American hornbeam	native	full sun, part shade, full shade	fine, medium, coarse	slow	tree	20	3–8	short	available	
Carya alba	Y	mockernut hickory	native	part shade, full shade	fine, medium, coarse	slow	tree	18–85	5–8	moderate	field collections only	prefers well drained soils, ridges, hillsides
Carya glabra	Y	pignut hickory	native	full sun, part shade	medium, coarse	slow	tree	30–80	5–9	moderate	contracting only	grows well in dry conditions; very drought tolerant
Carya ovata	Y	shagbark hickory	native	full sun, part shade, full shade	fine, medium, coarse	slow	tree	15–75	5–8	long	available	nuts can damage cars; do not put near streets
Celtis occidentalis	Y	common hackberry	native	full sun, part shade, full shade	fine, medium, coarse	rapid	tree, shrub	26–60	3–9	moderate	available	
Cornus alternifolia	Y	dogwood	native	full sun, part shade, full shade	medium	moderate	tree	25	3–8	moderate	no known source	
Cornus amomum	Y	silky dogwood	native	full sun, part shade, full shade	fine, medium, coarse	moderate	shrub	7–20	4–8	moderate	available	prefers moist soils
Cornus canadensis	Y	bunchberry dogwood	native	part shade, full shade	medium	slow	subshrub, shrub, herb	0.5	2–6	long	contracting only	prefers moist soils
Cornus racemosa	Y	gray dogwood	native	full sun, part shade, full shade	fine, medium	moderate	shrub	6–10	5–8	moderate	available	highly adaptable
Cornus sericea	Y	redosier dogwood	native	part shade	fine, medium, coarse	rapid	tree, shrub	7–10	2–7	moderate	available	naturally found near wetlands

Appendix 3 Cont'd

Species	Common name	Observed?	Invasive/ native	Sun amount	Soil texture	Growth rate	Growth habit	Size class (ft)	Hardiness (RI 5–7)	Lifespan	Commercial	Additional
Crataegus crus-galli	cockspur hawthorn	Y	native	full sun, partial shade	fine, medium, coarse	moderate	tree, shrub	30	3–7	long	available	used as ornamental
Crataegus phaenopyrum	Washington hawthorn	Y	native	full sun, part shade	fine, medium	moderate	tree, shrub	25–30	4–8	long	available	used as ornamental
Fagus grandifolia	American beech	Y	native	part shade, full shade	medium, coarse	slow	tree	30–80	3–9	long	available	
Fraxinus spp.	ash	N	native	full sun, partial shade	medium, coarse	rapid	tree	30	4–9	moderate	available	
Gaylussacia spp.	huckleberry	N	native	full sun, partial shade	medium, coarse	rapid	shrub	3–6	3–8	moderate	available	
Ilex glabra	gray inkberry	Y	native	full sun, part shade, full shade	fine, medium, coarse	slow	shrub	5 to	4–9	long	available	male and female specific plants
Ilex laevigata	gray smooth winterberry	Y	native	part shade, full shade	fine	moderate	shrub	10–12	5–8	short	available	prefers woodland swamps
Ilex verticillata	common winterberry	Y	native	full sun, part shade, full shade	fine, medium	moderate	tree, shrub	6–10	3–9	moderate	available	
Juglans cinerea	butternut	Y	native	full sun	medium, coarse	rapid	tree	20–80	3–7	short	available	
Juglans nigra	black walnut	Y	native	full sun, part shade	medium	rapid	tree	35–100	4–9	moderate	available	
Juniperus communis	common juniper	N	native	full sun, part shade	medium, coarse	slow	shrub	4	4–9	long	available	
Malus spp.	crabapple	N	native	full sun	medium, coarse	moderate	tree, shrub	30	4–9	long	available	
Morus rubra	red mulberry	Y	native	full sun, part shade, full shade	sand, loam, clay	moderate	tree, shrub	12–36	5–9	long (120 yr)	available	endangered in CT, MA
Myrica pensylvanica	northern bayberry	Y	native	full sun, part shade, full shade	medium, coarse	slow	tree, shrub	9–12	3–6	long	available	nitrogen fixing; male and female plants separate; berries only on F
Nyssa sylvatica	marshall blackgum	Y	native	full sun, part shade, full shade	medium, coarse	moderate	tree	30–95	5–9	moderate	available	wetland indicator

Appendix 3 Cont'd

Species	Observed?	Common name	Invasive/native	Sun amount	Soil texture	Growth rate	Growth habit	Size class (ft)	Hardiness (RI 5–7)	Lifespan	Commercial	Additional
Parthenocissus quinquefolia	Y	Virginia creeper	native	part shade, full shade	fine, medium	rapid	vine	1	3–10	moderate	available	
Picea glauca	Y	white spruce	native	full sun, part shade, full shade		moderate	tree	18–20	5–7	long	available	early seral
Picea pungens	Y	blue spruce	introduced	part sun, part shade	medium, coarse	slow	tree	20–100	4–7	long	available	
Picea rubens	Y	red spruce	native	full sun, full shade	fine, medium, coarse	slow	tree	30–100	5–7	moderate	available	
Pinus rigida	Y	pitch pine	native	full sun	medium, coarse	rapid	tree	20–80	4–7	moderate	available	inhabits coast
Pinus strobus	Y	eastern white pine	native	full sun, part shade, full shade	medium, coarse	rapid	tree	20–80	3–7	moderate	available	requires early weed control
Pinus sylvestris	Y	scotch pine	introduced	full sun	medium, coarse	rapid	tree	30–110	3–8	moderate	available	
Populus deltoides	Y	eastern cottonwood	native	full sun, part shade, full shade	fine, medium, coarse	rapid	tree	80–190	3–9	short	available	
Populus grandidentata	Y	bigtooth aspen	native	full sun, part shade, full shade	medium, coarse	rapid	tree	40–65	3–9	short	available	
Populus tremuloides	Y	quaking aspen	native	full sun, part shade, full shade	fine, medium, coarse	rapid	tree	40–65	1–8	short	available	
Prunus pensylvanica	Y	pin cherry, fire cherry	native	full sun	fine, medium, coarse	rapid	shrub, tree	25–30	3–8	short	available	
Prunus serotina	Y	black cherry, rum cherry	native	full sun, part shade, full shade	medium, coarse	rapid	shrub, tree	40–80	4–9	moderate	available	
Prunus virginiana	Y	chokecherry	native	full sun, part shade, full shade	fine, medium, coarse	rapid	shrub, tree	15–25	2–7	short	available	
Quercus alba	Y	northern white oak	native	full sun, part shade, full shade	medium, coarse	slow	tree	25–100	3–8	long	available	

Appendix 3 Cont'd

Species	Observed?	Common name	Invasive/ native	Sun amount	Soil texture	Growth rate	Growth habit	Size class (ft)	Hardiness (RI 5–7)	Lifespan	Commercial	Additional
Quercus coccinea	Y	scarlet oak	native	full sun	medium, coarse	rapid	tree	30–70	4–8	long	no known source	
Quercus palustris	Y	pin oak	native	full sun, part shade, full shade	fine, medium	rapid	tree	40–100	4–8	moderate	available	
Quercus rubra	Y	northern red oak	native	full sun, part shade	fine, medium, coarse	moderate	tree	36–81	4–8	long	available	
Quercus velutina	Y	black oak	native	full sun, part shade	fine, medium, coarse	moderate	tree	25–80	4–9	moderate	available	
Rhus hirta	Y	staghorn sumac	native	full sun	medium, coarse	rapid	shrub, tree	30	4–7	short	available	
Ribes americanum	Y	American black currant	native	full shade, part shade, full sun	fine, medium, coarse	rapid	shrub	15–30	3–6	short	available	
Rosa carolina	Y	Carolina rose	native	part shade, full sun	medium, coarse	moderate	subshrub	5	5–8	moderate	available	disturbed areas, roadside
Rosa virginiana	Y	Virginia rose	native	part shade, full sun	medium, coarse	moderate	subshrub	6	4–7	moderate	available	
Rubus allegheniesis	Y	Allegheny blackberry	native	full, partial	fine, medium, coarse	rapid	thicket	1–6	6–9	short	available	
Rubus flagellaris	Y	common dewberry	native	full, partial	clay, loam, sand, rocky	rapid	thicket, vine	3	6–9	short	available	threatened in Indiana
Rubus idaeus	Y	American red raspberry	native	full sun	fine, medium, coarse	moderate	subshrub	6–9	5–9	short	available	
Rubus occidentalis	Y	black raspberry	native	part shade, full sun	fine, medium	rapid	subshrub	5–6	4–9	short	available	
Rubus odoratus	Y	purple flowering raspberry	native	part shade, full sun	fine, medium, coarse	rapid	subshrub	5	3–8	short	no known commercial source	

Appendix 3 Cont'd

Species	Observed?	Common name	Invasive/native	Sun amount	Soil texture	Growth rate	Growth habit	Size class (ft)	Hardiness (RI 5-7)	Lifespan	Commercial	Additional
Sambucus canadensis	Y	common elderberry	native	part shade, full sun	medium	rapid	shrub, tree	7	4–9	moderate	available	
Sambucus racemosa	Y	red elderberry	native	part shade, full sun	medium, coarse	moderate	shrub, tree	10–20	1–5	moderate	available	historical; early seral; inhabits riverbanks
Sorbus americana	N	American mountain ash	native	full sun	fine, medium, coarse	moderate	shrub, tree	30	3–8	moderate	available	
Spiraea spp.	N	meadowsweet	native	part shade, full sun	fine, medium, coarse	rapid	shrub	4	4–9	long	available	
Symphoricarpos spp.	N	snowberry	native	part shade, full sun	fine, medium, coarse	rapid	shrub	4	4–9	long	available	grows well in urban areas
Ulmus americana	Y	American Elm	native	part shade, full sun	fine, medium, coarse	rapid	tree	50–120	3–9	moderate	available	
Vaccinium angustifolium	Y	low bush blueberry	native	full shade, part shade, full sun	fine, medium, coarse	moderate	subshrub, shrub	1–2	2–5	moderate	available	
Vaccinium corymbosum	Y	high bush blueberry	native	full shade, full sun	fine, medium, coarse	moderate	shrub	12	6–10	moderate	available	
Viburnum dentatum	Y	southern arrowwood	native	part shade, full sun	medium, coarse	moderate	shrub	3–9	5–7	moderate	available	
Viburnum lentago	Y	nannyberry	native	part shade, full sun	fine, medium	slow	shrub, tree	28	5–7	long	available	

 **EPA**

United States
Environmental Protection
Agency

Office of Research and Development
National Health and Environmental
  Effects Research Laboratory
Atlantic Ecology Division
Narragansett, RI 02882

Official Business
Penalty for Private use
$300

www.ingramcontent.com/pod-product-compliance
Lightning Source LLC
Chambersburg PA
CBHW081619170526
45166CB00009B/3035